AIRCHECK

The Story of Top 40 Radio in San Diego

DAVID LEONARD

On the Cover:
The name Shotgun Tom Kelly is synonymous with radio in San Diego. Shotgun is shown early in his career in the studio of KGB AM. This was the age of Bill Drake's Boss Radio when dialogue was tightly scripted. Liner notes stand above the console for air staff to follow. In the rear is a reminder from the program director to promote the weekend contest. The KFSD (now KOGO) Radio antenna is shown extending above the U.S. Grant Hotel as it appeared in 1941.

First Edition Limited to 1000 Copies
2003

To Kim

*Whose Support, Talents and Insights
were essential*

Preface

This story was made possible by the Internet. The resources of the Internet include a wide selection of radio stations tributes and histories, repositories of preserved Top 40 broadcasts, personal and station web pages, directories of station staffs, and on-line auctions offering radio memorabilia. It was the on-line auctions that placed me in touch with radio insiders who could explain the background behind the material I felt compelled to purchase. Using all of these resources, I began to organize DJ line-ups and notes for personal use.

Carol Craig used to work at KCBQ and hosts a web page celebrating the station. With her encouragement, and by probing the encyclopedic minds of DJ Gene Knight and engineer George Junak, I became a student of how Top 40 radio functioned. Only after absorbing months of tutoring by them, and exhaustive searches of information on the internet, did I dare begin a series of contacts with air personalities and management staff.

As my notes grew in volume, personal stories began to emerge that would challenge the imaginations of the best Hollywood script writers. I started with the early 1970s, intending to focus on the KCBQ/KGB radio war of 1971. I contacted people all over the country who were in San Diego radio at that time. The story was to end at the time I left the area in 1972. But a KCBQ history web site provided the names of its on-air staff of the 1950s, so I leaped back to that era. Author Bill Earl, (*When Radio Was Boss, Dream-House*), had poured through radio logs from the *San Diego Union* that provided a complete set of line-ups from 1955 to 1965. Bill provided me with that information, so then it was a matter of finishing the story to 1972. But many events started in 1972 that concluded in 1976. After some consideration, I decided to cover the complete Top 40 run of all AM stations.

I continued to contact DJs wherever I could find them. Interviews were mostly by e-mail. Occasionally, a telephone interview was conducted. Shotgun Tom Kelly hosted a mini-reunion that served as one face-to-face session I had with local DJs. A lunch with Gene Knight was the only other. Long-time San Diego personalities certainly have great stories to share. But some of the best were from people who were in the market for only weeks.

Some personalities expressed a little annoyance over the accuracy of information I was collecting. This led to a tip that I should be more resolute in contacts with Program Directors who could offer expanded insights. The PD contacts linked these individual stories into an eloquent, if not very direct, explanation of how things happened and why.

All the information I could collect from or about each personality, with a reasonable degree of accuracy, is included in this compendium of pop culture. Information about personal lives, unrelated to radio, has remained private. Personal information that was reflective of life in the industry is included. Some personalities died before contact could be made. Some could not be found. So, as with any history, this is a story told with the best resources available. The stories unfold with the years and the station locations each person worked.

Contents

Introduction 1
    Overview of KCBQ 3
    Overview of KGB 5
    Overview of KDEO 7
    Overview of XEAK 8
    How the Stations Rated 9

KCBQ Radio AM 1170 11
    Top 40 Radio Comes to San Diego 11
    1955-1965: The Hitparaders 12
    1965-1967: Blood Is Thicker Than Water 17
    1968: Texas Mafia Snuffs Boss Radio 26
    1969-1970: Allyn Tyme 30
    1971: The New Q 35
    1976-1980: Toy Soldiers to Singing Cowboys 44
    Images of KCBQ 49

KGB Radio AM 1360 55
    Inaugural Top 40 Format and Staff 55
    Boss Radio 56
    Buzz Radio 63
    KGB Radio 67
    Recycled Rock 69
    The 13K Era 71
    Images of KGB 75

KDEO Raydeo AM 910 79
    Images of KDEO 88

XEAK Radio AM 690 91
    Images of XEAK 97

Coverage Maps 99

Glossary 109

Index 111

Acknowledgements

This is a story as told to the author by the participants. The participants were asked to recall details of events that had occurred 30 to 50 years prior, and most did with amazing clarity and conviction. The San Diego radio market was an extremely competitive environment in the 1960s and some of that competitive spirit continued to be exhibited during personal interviews. The prevailing opinion is that the "Silver Age of Radio", as Bill Earl coined it, occurred during this era. Many considered their San Diego tenures to be the highlight of their careers. A big thank you to the following folks that contributed moments or hours of time toward this story:

Gary Allyn
Chris Bailey
John Barcroft
Kevin Barrett
Harry Birrell (Jerry Walker)
Jerry G. Bishop
B. Baily Brown
Jim Carson
Gentleman Jim Carter
Noel Confer
Paul Christy (Johnny Mitchell)
Chuck Cooper
Carol Craig
Chuck Daugherty
Jerry Davis
Bill Earl
Phil Flowers
Don 'Sonny' Fox
Chuck Geiger
Dean Goss
Mel Hall
Jack Hayes
Johnny Hayes
Drew Harold (Bobby Noonan)
Doug Herman
Bill Hermanson (Cap'n Billy)

Sie Holliday
George Junak
Barry Kaye
Shotgun Tom Kelly
Gene Knight
Jim LaMarca
Tom Leland (Gene West)
Dave London
Pat Maestro
'Happy Hare' Martin
Danny Martinez
Jack McCoy
Roger W. Morgan
Jim Nelson (Tony Evans)
Bobby Ocean
Ron Peer
Sonny Jim Price
Jimmy Rabbitt
Rich Brother Robbin
John Dakin (Chuck Roberts)
Neilson Ross
Jim Ruddle
Tommy Sarmiento/Tommy Lee
Tom Schaeffer
Lee Shoblom
Jim Scott (Stan Walker)

Big Mike Scott
Bob Shannon
Michael Spears (Mark Richards)
Frank Thompson
Stan Torgerson
Charlie Tuna
Jack Vincent
Bill Wade
Beau Weaver
Sonny West
Brian White
Ray Willes
Johnny Williams
Danny Wright

PHOTO CREDITS: Bill Earl from his private collection and book *When Radio Was Boss;* Carol Craig from original work as a KCBQ photographer; from others among the list above, particularly Gene Knight, Shotgun Tom Kelly, and Chuck Daugherty; and from the weekly surveys produced by all stations. Special thanks to Millie Cessna and A to Z Printing for their graphic design artistry.

Introduction

For over 70 years, radio broadcasting had been a prominent industry in San Diego, and the San Diego market has been prominent in the radio broadcasting industry. KOGO radio was licensed on June 2, 1925 as KFVW, went on the air in March 1926, and changed their calls to KFSD (First in San Diego) in April 1926. Even in these early days of radio, facts were not allowed to get in the way of a slick marketing plan, because what is known today as KPOP was licensed in July 1922 as KFBC, well before KFSD. KFBC changed its calls in 1928 to KGB. Long before wartime operations elevated the national prominence of San Diego, radio network executives had targeted San Diego as a market to establish affiliate stations. Such wisdom paid off as San Diego radio stations flourished during the Great Depression. During the 1940s, all major radio networks established affiliates in San Diego that featured live audiences, orchestras, and large dance floors. The affiliates were KFSD (now KOGO) for NBC, KGB (now KPOP) for Mutual-Don Lee then for ABC, KSDJ (now KCBQ) for CBS, and KFMB for ABC then CBS.

As radio stations lost their audience to television in the early 1950s, air time had to be re-programmed to attract and hold an audience while capturing advertisers that would maintain profitability. Hoyle Dempsey writes that "Lifestyles were changing, portable radios were becoming common, people were spending more time in their cars, and radio could be the perfect accompanying entertainment, not just for part of the day, but for the most part of the day."

Several broadcast companies started experimenting with programming to attract a broad audience. Refinements to this process never really stopped. Among these broadcast companies was Bartell Media, who bought KCBQ in 1955. Bartell was among the early innovators of the program clock that synchronized (and repeated) news, weather, announcements, advertising, and jingles around a careful rotation of songs. Commonly known as Top 40 radio, for the number of songs on the play list rotation, it constituted formula radio to achieve listener loyalty. In its earliest form, Top 40 formats were designed to reach all age groups by carefully delegating the type of music and information aired over the course of the day. But the British Invasion in 1964 identified an energized demographic base with unlimited economic potential. This changed Top 40 radio over night to a medium aimed at teens and young adults.

The coveted network affiliations of the 1940s that produced the majority of programs for airing were not compatible with the Top 40 formats emerging in the 1950s. The Bartell Group understood this and dropped the CBS affiliation when they bought KCBQ. It took longer for Brown Broadcasting to understand this before they dropped their ABC affiliation for KGB in 1964. But dropping network affiliations also meant losing revenue from network advertisers as well as public service programs that satisfied FCC broadcast licensing requirements.

Stations adopting Top 40 formats attracted large ratings increases that network programming had not seen in years. Advertisers were quick to sign on with winners. It would be an awkward transition to public service programs, but the FCC had no restrictions on the placement of public service programs and time slots were carefully selected that tended to have the fewest listeners. Excellent documentaries and contemporary talk radio formats emerged from these public service obligations.

Top 40 radio also heightened the now-lost concept of on-air personalities who worked the

INTRODUCTION

control boards, spun the discs, punched up the ads and jingles, and engaged listeners in conversation. To some, they were companions during the day, to others they were in-the-know celebrities, to teen age girls they were icons. These disc jockeys were generally successful because they were clever, funny, and upbeat. Some tended to prefer controversial jocks. Others gravitated toward appearance (as seen from weekly surveys), or those that came across smooth or sexy on the radio. Jocks who were knowledgeable about the music or the latest social or cultural movements would enjoy broad popularity.

To many of us, the job seemed easy. Just sit there and talk, read ads, and play records. But for a station to gain audience ratings, it needed continuity in its programming. This was achieved by the program clock where music, news, announcements, contests, and ads were synchronized to designated parts of the hour. Every station had a program director to maintain conformity to the format clock and tempo of the broadcast. Disc jockeys had to conform to the format, yet appear to be spontaneous and interesting in order to hold an audience. They had to draw enough audience to attract and maintain advertisers. But they had to limit ads and play enough music to satisfy the listener. The disc jockey needed to do all those things while operating a control board, punching up or reading ads, synchronizing cues with the station engineer (if there was one), loading carts, and making announcements or introductions as a lead in to a record while hitting the post just as the vocals began. The job wasn't that easy.

Disc jockeys worked in shifts that were scheduled to match certain types of jocks with the types of listeners through the day and night. During the 1950s, shifts were split where a jock would be on for two hours, off for two hours, then return for two or three more before ending his day. However, the normal schedule was divided among three or four hour shifts through out the day. The characteristics of the various shifts followed along this general reasoning:

- Morning drive was almost always 6-9 AM. A desirable jock in the shift would be witty, topical, upbeat, and energetic. The audience were people starting their day in their homes or in their cars going to work or school. This shift was important to attract listeners at the beginning of the day in the hopes of keeping them tuned in.

- Mid days were almost always 9-Noon. A desirable jock sounded mature and natural on the air. He might sound witty, sexy and smooth in his delivery. Programming was aimed primarily at housewives.

- Afternoons were usually Noon to 3 or 4. Pace and tempo were the order here. Listeners would be busy going about their day and the desirable jock would help set a pace with music and announcements.

- Afternoon drive varied between the hours of 3 to 7 PM. A desirable jock would be knowledgeable, witty, and 'hip' for rush hour drivers and after school listeners. This was a highly competitive shift that attracted higher advertising dollars and audiences. Talent and experience were key assets.

- Evenings were usually between 4 and 9 PM. This was party time and required big energetic voices to excite the listener about the music and related events, such as concert information. Programming was typically aimed at teenagers and young adults.

INTRODUCTION

- Nights were typically from 8 to Midnight. The pace got slightly slower, the music slightly softer, or took on a lyrical edge, as the day wound down and the audience was slightly older. Some of the greatest programming in Top 40 radio occurred during these hours in the form of non charted underground or obscure album cuts that featured the best music instead of the hits. Talk radio made an entry from time to time during this shift, often to meet FCC 'public benefit' requirements.

- Overnights were Midnight to 6 AM. This audience would be the convenient store operators, night watchmen, graveyard shift employees, and other nocturnals. This was the longest daily shift and the jock needed to sound fresh and consistent from beginning to end.

Nearly every Top 40 radio station in San Diego followed this regime in one form or another. This story focuses on four stations that successfully aired Top 40 programming during various parts of the 1960s decade. All of these stations contributed toward the development of air talent, radio management, formats, and contests. The station that introduced the Top 40 genre to San Diego and remained an industry powerhouse for decades was KCBQ.

Overview of KCBQ

KCBQ was licensed in 1946 as KSDJ reflecting its ownership by Clinton McKinnon's *San Diego Journal* newspaper. McKinnon also served two terms in Congress representing San Diego from 1948-52. The station operated at 5000 watts and 1170 kilocycles as an affiliate of the prestigious and dominant CBS radio network. The station conducted live broadcasts and transmitted network programs from its studios at the Journal Building on 5th and Ash.

KSDJ was sold to dentist Charles Elliot Salek in 1949. He applied for and received new call letters KCBS to reflect the network affiliation. CBS wanted to use those calls for the affiliate they owned and operated in San Francisco, so they bought the rights from Salek. He then agreed to alter the calls to KCBQ- for Quality. Salek was actively involved in station operations, but knew very little about radio. But by surrounding himself with excellent broadcasters, the station flourished with network programming and live studio entertainment through the early 1950s. Salek also relocated the studios to the Imig Manor Hotel at El Cajon Blvd. at Texas St. in 1951. At the time, the transmitter was located on El Cajon Blvd., adjacent to the Campus Drive-In Theater.

By 1954, KCBQ had become an ABC network affiliate offering a mix of network and local shows. A typical weekday line-up featured *Morning Chapel Hour* at 7:00 AM, *Breakfast Club* morning show at 8:00, news with Chet Huntley at 9:00, and several radio soaps until 1:00 PM. *Bandstand and Grandstand* aired until 5:00 PM followed by news at 6:00. The night schedule was dominated by *Moods in Music. Zanz's Private Line* was the overnight show.

As traditional radio programming was losing audience and profitability because of television, KCBQ was sold to Bartell Family Radio in 1955. The Bartell media empire included ownership in publishing and eight radio stations. These stations included WOKY in Milwaukee, KYA in San Francisco, WYDE in Birmingham, WAKE in Atlanta, (known as WYDE a-WAKE radio of the south), WILD in Boston, and KRUX in Pheonix). The CBS affiliation had returned in 1955, and was immediately dropped as the Top 40 music format was adopted. The station transmitter was located

INTRODUCTION

at 9416 Mission Gorge Road in Santee, east of San Diego. KCBQ rapidly moved to a ratings leader in San Diego.

The station outgrew their hotel location and relocated to new facilities at 7th St. and Ash in downtown San Diego after their lease expired with the hotel in July 1957. The station would remain at this location for the next 11 years. The station received FCC approval in 1958 to begin broadcasting at 50,000 daytime watts, while retaining 5000 watts at night

1968 was a banner year for the station. The studios were relocated to larger facilities in Santee at the transmitter site where the station would remain for the next 30 years.

By 1980, KCBQ was one of the last AM stations to drop the Top 40 format. From that point on, the station simulcasted on AM and FM bands, and was shuffled as a property from ownership to ownership using Country & Western, followed by Oldies, formats.

Bartell Media sold their radio chain in 1977 to Charter Broadcasting. The station was sold to Wheeler Broadcasting in 1980. Wheeler moved KCBQ-AM into a Country & Western format in 1980. Simulcasting with KCBQ-FM began in 1981. They were making gains on KSON in 1984 when the death of then-owner John Bayliss eclipsed the broadcast side of the business. Wheeler Broadcasting sold the station back to Charter Broadcasting and the KCBQ calls were de-emphasized. Charter then sold to Infinity, who brought back the KCBQ calls, then sold to Eric Chandler Ltd. who turned the format over to Oldies music in 1986. Sonny West and Bill Moffitt remained with the station through most of the decade. Eric Chandler Ltd sold to Adams Communications in 1989. The station pioneered a modern oldies concept in 1993 bringing back air staff from the 1970s and 1980s. By 1996 Compass Radio Group owned the station, who sold it to PAR Broadcasting. PAR was purchased by Jacor in June 1997, which already owned KPOP, (the former KGB station), and other San Diego stations. Pursuant to FCC market ownership regulations, Jacor had to sell KCBQ. Salem purchased the station from JS Broadcasting in 1999 and took it from oldies to a talk radio format.

Due to equipment deterioration, structural problems, area development, and interference by other stations, KCBQ was unable to maintain their licensed night power of 5000 watts and decreased the power output in the mid 1990s to 1500 watts.

Salem Broadcasting had purchased the station in 1999 and relocated the studios to University City to share with sister station KPRZ AM 1210. The format is talk radio dominated by conservative hosts.

The site of the transmitter and former studio in Santee was sold for development as a home improvement center. The application to relocate the transmitter to Muth Valley was denied by the County of San Diego in 2003. Arrangements were then made to have the station temporarily move its transmitting facilities and di-plex with another station. This move results in a significant reduction in the daytime signal strength. That other station is KPOP 1360, formerly the home of KGB.

The self-proclaimed 'Legendary Station KCBQ' is well deserved based on the rich and colorful history it has developed for decades. A factor that made KCBQ reach higher and try harder was the considerable competition generated by KGB radio during the years it served as Bill Drake's first *Boss Radio* station.

INTRODUCTION

Overview of KGB

As the oldest continuing radio station in the San Diego Market, KGB has been the subject of previous writings. *Radio Station KGB and the Development of Commercial Radio in San Diego,* by Marie Breen Crane and published in *The Journal of San Diego History* in 1977, offers an excellent history of the station leading up to the Top 40 era. *Radio's Golden Era, Radio And Wireless As They Used To Be,* by Alice Brannigan and published in *Popular Communications* in October 1990 summarizes the development of KGB radio. *When Radio Was Boss,* by Bill Earl is a fun scrapbook focusing on California, Arizona, and Mexican Top 40 stations, including KGB and KCBQ. Numerous web sites celebrate the events surrounding KGB as the catalyst for the successful Boss Radio format. Another substantial resource is *When the Music Stops It's News.* This production aired during the final days of the 13K format and focused on the years 1959 to 1982.

The station began as a homemade experiment. W.K. Azbill was granted a license on July 14,1922 to begin broadcasting as KFBC, 'The Normal Heights Station', from his residence at 5038 Cliff Pl. (The residence has been replaced by newer homes on the site). The station initially aired at 10 watts of power, but since licenses were issued by the Department of Commerce every three months, power and frequency changes occurred often. Azbill is credited as being lucky to have survived the high mortality of small stations during this period to the point that he was able to secure 50 watts of power by 1927, when the station operated at 1210 kilocycles. This license was assigned to Dr. Arthur Wells Yale in 1927.

Pickwick Broadcasting Corporation bought the station in 1928 and installed George Bowles as Vice President and Manager of the station. The call letters were changed to reflect his name as KGB. Under the Pickwick ownership, the station began operating at 1330 kilocycles with studios and dual transmitter towers in the Pickwick Hotel located at Broadway and First. Stations used a variety of slogans to promote their identity. Among those KGB used during this time were "The Sunshine State of California" and "Music for the Sick".

Don Lee had amassed a substantial fortune from Cadillac and LaSalle automobile dealerships and decided to get into broadcasting. He bought KFRC San Francisco in 1926 and KHJ Los Angeles in 1927. In 1929, CBS sought affiliate stations west of the Rockies and an affiliate agreement was quickly negotiated with Don Lee that created the Don Lee-Columbia network on July 16, 1929. The Vice President of Don Lee Broadcasting was Willet H. Brown. KGB was purchased by Lee in 1931 and added to the network. After Don Lee died suddenly of heart failure on August 30,1934, the network was run by his son and sole heir Thomas S. Lee. He assigned the KGB license to station manager Marion Harris.

Art Linkletter broke into broadcasting as a KGB announcer in 1934. He moved up the ranks to Chief Announcer/Program Director by the time he left in 1942. TV writer Larry Rhine began his professional career at KGB in 1934 and went on to become a screenwriter and TV comedy writer for such shows as 'Red Skelton', 'Bob Hope', 'Mr. Ed', 'Here's Lucy', and 'All in the Family'. Meanwhile, the hasty acquisition of Don Lee stations had led to friction between CBS and Don Lee Broadcasting. When CBS acquired KNX Los Angeles in 1936, the affiliation between the two was terminated. Don Lee Broadcasting was merged with Mutual Broadcasting on December 29, 1936. KGB began operating at 1360 kilocycles in 1942.

INTRODUCTION

By 1949, KGB was operating at 1000 watts. Following the suicide of Thomas Lee in 1950, the Don Lee network was sold to General Teleradio, a division of General Tire and Rubber Company, in 1951. A new studio for KGB was constructed at 4141 Pacific Coast Highway in 1955. General Teleradio had acquired RKO Radio Pictures from Howard Hughes in 1954 to form RKO General. The station was sold to Marion Harris at this time who increased the output to 5000 watts. The station affiliated with the ABC network during the late 1950s and installed Bob Regan as Program Director. He was later at CBS affiliate KFMB-TV as a newsman. The station transmitter was located at 2302 - 52nd St., at Kalmia.

Former Don Lee Broadcasting VP Willet Brown formed Brown Broadcasting Company with his son Mike and purchased the station in 1961. Willet was involved in the Columbia Network merger in 1929, the Mutual Broadcasting merger in 1936, and was likely involved in subsequent transactions with General Teleradio and RKO Radio pictures. He owned a cadillac dealership, a yaught, and his own Greyhound bus that was a familiar site at the studio. He was a man accustomed to success.

By 1963, KGB was attempting to become a Top 40 station while being a major network affiliate. The two were not compatible. As the station developed a middle-of-the road (MOR) music format around youth, ABC was providing news feed aimed at adults. The Browns began the search for a strong proven programmer to resolve this conflict. They initially sought out the programmer of KMEN in San Bernardino, but didn't find who they were looking for. (Ron Jacobs had already moved on to make history at KMAK in Fresno). His rival, Gene Chenault of KYNO Fresno, was trying to branch out in his new radio consulting business. Chenault was hired as the station program consultant after meeting with the Browns. At the same time, Chenault was creating a partnership with Bill Drake and had secured a consulting contract for the radio side of RKO General, owner of KHJ in Los Angeles.

Chenault and Drake had experimented with a faster paced programming format in Fresno. They dropped the ABC affiliation and hired several DJs in 1964 from the Fresno market. After further experimentation, the format was perfected in 1965 as the *Boss Radio* concept was born. Despite being the first station programmed by Chenault and Drake, KGB struggled against local competition, (namely KCBQ), to enjoy the ratings dominance that the other stations they programmed, (KFRC in San Francisco, KHJ in Los Angeles, and KYNO in Fresno), had in their markets. Ironically, this program format network looked very similar to the network of stations owned by Don Lee Broadcasting.

By 1972, *Boss Radio* had been copied and refined by others. A format makeover was performed to create an album-oriented Recycled Rock format. KGB AM and FM began simulcasting as the station plunged into community-oriented projects. These would include the homegrown series of albums featuring local bands with proceeds donated to the United Way, an annual Skyshow on July 4th, and the introduction of the KGB Chicken. Despite many award winning productions, the station was losing market share until a new format with a new service mark was introduced in 1979.

KGB debuted the 13K service mark in October 1979 with a return to the traditional Top 40 format, but with a music list and personalities considerably more mellow than that of *Boss Radio* in order to appeal to an adult audience. The station returned to the top of the ratings and cut short a

INTRODUCTION

renaissance of Magic 91, formerly known as Raydeo KDEO. But the 1982 emergence of The Mighty 690 XETRA, formerly known as XEAK, brought a close to the 13K era.

The station leased its air time to Turner network in 1982 and was used as an experimental feed for Turner Network Headline News. The calls were changed to KCNN. This was not a profitable venture. Brown Broadcasting continued to own the station until 1997 when it was sold to Nationwide Insurance, who sold it to JACOR that same year.

Clear Channel Network now operates KPOP on this frequency that offers nostalgic sounds of the big band era, jazz, and pop songs anchored by heritage jocks who have deep roots in San Diego. Interestingly, these include Happy Hare who was at KCBQ when the Top 40 format was ushered in, and Jerry G. Bishop who was at KCBQ when the format was ushered out.

Overview of KDEO

Tullis and Hearne, owners of several California Top 40 radio stations, also had a San Diego presence in addition to the Bartell ownership network, and the Boss Radio format network during the 1960s.

Babcock Broadcasting received a license for a station at 200 East Main St. in El Cajon with the calls KBAB in the early 1950s. The station was affiliated with the Keystone Broadcasting System. The format offered a vignette of local programs in the morning for La Mesa, Lemon Grove, Chula Vista, and National City. A major league baseball game of the day was offered in the afternoon. Music was featured in afternoons with K-BAB Bandstand, and at nights with Club 910. Soon, a cluster of cabanas was fashioned into a studio at the Town and Country Hotel by 1958.

Dandy Broadcasting acquired the station in 1959 and inaugurated a Top 40 format. The Dandy network was operated by Ken Greenwood, Dick Harris, and Bob Sharon who all had backgrounds in sales. With a financial backer, they acquired WPEO in Peoria, IL, KLEO in Wichita, KS, and KQEO in Albuquerque, NM. They changed the calls from KBAB to KDEO in 1959 to conform to their station ownership trademark. The facility was relocated to 3911 Pacific Coast Highway on the second floor of the Machinist's Union Building. The problem with this location was the proximity to Lindbergh Field that created aircraft noise interference during broadcasts. The station finally located at 2262 Fletcher Parkway by 1962. The former studio now houses Fletcher Hills Printing. A 1000 watt transmitter building, with three antenna arrays, was built at North Magnolia and El Nopal in Santee. The building was painted bright yellow with red KDEO letters on the front facing the street. This site was demolished in the 1970s. Dandy had to liquidate its assets in 1960 when the financial backer died.

The station was acquired by Howard Tullis and John Hearne during the summer of 1960. Howard Tullis had a background in radio sales that included a heavy ad buy at station KLAC in Los Angeles. John Hearne had been the attorney for KLAC and they knew each other well. Roy Cordell was brought on as the Group Programming Director. Tullis and Hearne formed a network that included stations KAFY in Bakersfield, KFXM in San Bernardino, and KDUO in San Bernardino. Unlike their Bartell and Drake counterparts, the focus at the network level was on sales, not programming. However, Howard Tullis could listen to KFXM from his Los Angeles office and would frequently drive out to get involved in the station programming. Programming was

INTRODUCTION

usually the exclusive domain of the station program director. When consultations did occur among station PDs, the topic was normally promotions rather than music.

KDEO was the first to offer extended music sets, the first to broadcast the American Top 40, and the first to air a female DJ at a Top 40 station in the market. KDEO management had a good ear for talent and music, but could not retain their talent for long periods of time as often required to build an audience. The 1000 watt coverage area was also a problem for competing with 50,000 watt KCBQ and 5000 watt KGB.

Tullis and Hearne sold the station in late 1965 to Mort Hall, Mort Sidley, and Don Balsamo. Hall was the owner of KLAC and husband to actress Ruth Roman. Sidley was the sales manager at KLAC, and Balsamo was the sales manager of KHJ-TV. Faced with the rigorous competition from KCBQ and KGB, the station started changing formats in February 1966, to all requests, followed by oldies, then adult standards. KDEO was the first station to broadcast the American Top 40 countdown on July 3, 1970.

KDEO went to an album rock format in 1976. Lee Bartell returned to the San Diego radio market when he acquired KDEO in 1976. His son Richard became the General Manager. They quickly moved to drop the album format in favor of a revitalized, and heavily market researched, Top 40 format. They re-imaged the station by changing the call letters in 1977 to KMaJiC and adopting the 'Magic 91' slogan. Ratings spiked high enough to knock KCBQ out of its long-standing market lead briefly in 1977. Station B100 (KFMB-FM) was the ultimate beneficiary as the break from KCBQ dominance made it easier for listeners to adapt to the FM band.

Top 40 had run its course by 1980. The station was sold and the KMJC calls were re-christened as (King and Master Jesus Christ) to reflect a new age of Christian broadcasting in September 1980. The station is now owned by Family Radio Network and continues religious broadcasting under the calls of KECR.

Overview of XEAK

Radio stations need not be affiliated with a network at all. Blend a Mexican station with some Tulsa, Oklahoma entrepreneurs, and some creative methods to work around Mexican and American broadcast laws and you get XEAK. Americans could not own a Mexican station, but they could lease all of its air time. Such was the arrangement that Jim and Bob Harmon made with owner/partner George Rivera as they took over 5000 watt station XEAC in 1956.

The Harmons immediately set out to gain approval to increase power to 50,000 watts, build a new transmitter near the town of La Presa, and a new studio in the Mission Valley area of San Diego, and changed the calls to XEAK. Daytime jocks lived in Baja, CA and did live broadcasts at the transmitter under work permits and visas. Night jocks taped their shows in a studio behind the Mission Valley Inn for rebroadcast because live international broadcasts at night were not allowed.

The Harmons hired talent from the San Diego area or those they had known in Tulsa that remained in the San Diego radio market for decades. The station had a great run. Though it never led the ratings in San Diego, it did lead from time to time in the Los Angeles market.

The station was sold to Gordon McLendon in May 1961 and became the first 24 hour all-news format. Mexican broadcast requirements for stations in this region mandated that calls begin with 'XE'. Therefore, XEAK became XETRA, with a service mark of X-TRA, to promote the news function. The station calls had to be announced as, "XETRA Tijuana, Mexico" in Spanish. Fearing that listeners would think they were listening to a Mexican music format, the station followed the ID with an ad promoting Mexico tourism. This would make it appear that the Spanish they just heard was part of the ad.

On April 2,1968, the station dropped all news and changed to Beautiful Music featuring the sounds of Ray Conniff, Ray Charles Singers, Nat King Cole, and others.

The Mighty 690 Top 40 format returned in 1980 and climbed to the top of the ratings by 1982 when it knocked out the KGB successor 13K format by using a tight play list of 'power hits'. But FM stations were now the preferred choice by Top 40 listeners, so the station went to X-TRA Gold in 1984 playing oldies records. A talk radio format was begun in 1988.

X-TRA Sports was inaugurated in 1990. The staff worked in San Diego with a satellite relay to the transmitter, now located at Rosarita Beach in Baja. Jacor purchased the station in 1996 and formed a network of Southern California stations. The transmitter was relocated to a site seven miles south of the cluster of transmitters below Rosarita Beach in 1998. The studio has been relocated to Los Angeles.

How the Stations Rated

The success of any radio station is measured by ratings. Ratings influence the rate advertisers pay for air time and how managers will make programming decisions. As in the case of television, consumers randomly received diaries in the mail to track the stations they listened to. They would return them to a rating company, such as Hooper in the 1960s, or Arbitron later, who tabulated the data and published the ratings. Archival ratings data, especially for periods extending back decades, are not easy to find. But program directors have an amazing ability to recall the performance of their station clearly.

The question about which station was best at any given time is the choice of the individual listener. But those stations that received the most diary responses during the seasonal sweeps, (and highest ratings), have been researched through interviews and limited published materials. KCBQ inaugurated the Top 40 format in the San Diego market in December 1955 under PD Allen Heacock, and was the ratings leader without interruption until 1961, when KDEO, under PD's Mel Hall followed by Noel Confer, had its lone moment on top. KCBQ returned to the top of the ratings during the next sweep in Spring 1962.

The year 1965 brought about major changes to Top 40 radio based on multiple factors. That was the year the British invasion was in full force. What had begun musically with the Beatles a year earlier was now manifesting itself in our speech, clothing, and entertainment that challenged and redefined the culture, especially the culture of the Top 40 audience. Personalities on the air were becoming aged and stale to youths who preferred younger more 'hip' personalities and who came across more informed on current trends. The pace of change was getting faster and faster.

During this time, Bill Drake, with his partner Gene Chenault, found a means to program at a faster pace, play more music, turnover more songs, and reduce chatter. This concept caught on

INTRODUCTION

with the Top 40 audience and propelled Drake's KGB to the top of the ratings in Fall 1965.

Program innovations and talent were responsible for KCBQ returning to the top, under PD Mike Scott, in Spring 1968, and for KGB doing the same under PD Buzz Bennett in Summer 1969. KCBQ took the lead back in Fall 1970 under PD Gary Allyn and stayed on top under numerous PDs until 1977. After that, no Top 40 AM station was dominant. KMJC (formerly KDEO) spiked in the 1977 ratings. KGB (now known as 13K) spiked in 1981, and The Mighty 690 X-TRA (formerly XEAK) spiked in 1982. But the highly consulted, highly researched, and highly categorized FM stations had become established with listeners.

Top 40 AM stations in San Diego were engaged in intense competition that frequently resulted in formats, contests, rules, or policies having national implications. Management and air personality staff found that mention of a San Diego station on their resume, particularly KGB or KCBQ, was very beneficial for the advancement of their career. This story endeavors to let the story unfold in the words and expressions of the participants who were there.

KCBQ Radio AM 1170

Top 40 Radio Comes to San Diego

The Bartell Group was among the first to offer successful independent programming without affiliations with a major network. They were among the first to create the traditional broadcast sound anchored by a program clock to set the tempo featuring music, local news, fast-talking DJs, public service announcements, jingles, contests, special promotions, and of course, the 40 most popular songs of a given time period.

The Bartell Group, (Bartell Media after 1965), was a family operation based in Milwaukee, WI involving brothers David, Lee, Melvin, and Gerald who served on the board of directors. David and Lee rotated turns as President. Their sister Rosa Bartell Evans and her husband Ralph were also closely involved with construction and operations. Lee Bartell was installed as President and General Manager of KCBQ in 1955, and successfully held that role into 1965. He is described as being 'the master of the ship.' He did the hiring and firing and would preside over staff meetings to lay out how he wanted things done. His ongoing theme was always "Brighten up and tighten up." He kept a residence in San Diego, but was based in New York. Lee would spend time each month in San Diego managing his assets. This 5'4" pipe smoking little general had been a successful corporate lawyer. His ambitions eventually led him to buy land in Mission Valley and develop hotels. He was the station commander through 1967.

Phyllis Sandin held the title of Music Director, but was far more than that. Her service began in 1955 as a secretary or administrative assistant to Lee Bartell. No one had the trust or devotion from Lee Bartell that she did. Their relationship is described as being very close, closer than his family. She was an older woman with white hair, rough skin, and seemed to drag a leg as she shuffled along, and who dressed very fashionably. She was a tireless employee who could be found at the station as early as 6:00 AM and as late as 10:00 PM. Sometimes she just spent the night at the station in Lee's office.

In her role as Music Director, she selected the records to be played on the air. She also selected the DJ picks for future hits. DJs had no input on the choices that appeared under their name on the weekly survey. By 1965, she had sufficient authority to limit the role of several DJ/Program Directors to that of program managers with little or no power over music or personnel decisions. By this time, Bill Drake had introduced *Boss Radio* to the San Diego Top 40 market, against which she was no match.

The third level of authority rested with the Program Director (PD). KCBQ PDs were relegated to the task of implementing management policy rather than to create and innovate. With the exception of the first PD, Allen Heacock, all others were selected from among the DJ staff and found the experience to be frustrating.

This management structure began to change in 1965. The Bartell Group acquired a publishing company, then reorganized as Bartell Media. Meanwhile, Boss Radio was catching on at KGB that caused Lee to panic and begin a string of poor decisions. The Bartell Board of Directors insisted that Lee Bartell return to New York to serve as President. He designated Phyllis Sandin to run the station. After the station lost market share and sales revenue, the Board insisted that Lee hire a general manager to run the station. Otherwise the board would sell the station. Lee and Phyllis devised methods to maintain control for another year. But the combination of unlawful eavesdropping and further company mergers led to the removal of both from all station control in 1968.

George Wilson (George Cowell) was hired as Group Program Director who rode honcho over all Bartell radio stations. He was a complex man who was once led an orchestra, became a disc jockey then a savvy program director at Bartell station WOKY in Milwaukee. He had authority over all personnel of the stations Bartell Media owned. Having been a DJ, he understood that talented air staff could create energy and enthusiasm over the airwaves, yet he felt any DJ, and sometimes an entire air staff, were expendable. Whenever someone wondered about adding a new song early or playing a questionable oldie, he observed that 'its not what you DON'T play that kills you, it's what you DO play.' He had a good heart, as evidenced by many occasions he would take an entire office out to dinner at a swank restaurant or lend money to employees to help make ends meet. He was cold blooded in his seemingly carefree manner of firing people. He found pleasure in heavy drinking and gambling. He had a commanding presence when he entered a room. Based in Milwaukee, Wilson came on the scene in 1968 and was a potent force in Bartell radio operations. He was promoted to President of Bartell radio 1974 and hired Jerry Clifton to succeed him as Group PD.

Next in line was the station General Manager. The GM oversaw finance, sales, and promotion matters. He could make personnel decisions involving the program director with the consent of the Group PD. There was frequent turnover in this position until the 1970s when Russ Wittberger arrived and remained for many years.

The Program Director took on substantial authority over music and air staff after 1968. With the exceptions of Mike Scott, Gary Allyn and Rich Brother Robbin, most program directors were not DJs and tended to focus more on the production aspects of broadcasts. Allyn and Robbin were again the exceptions. Their focus was on the music and had others handle production work. The new structure gave the program director hiring and firing authority over the DJs, just as most stations had been doing for years.

This story covers a 25-year period, from 1955 to 1980, when KCBQ offered a Top 40 format on the AM dial at 1170 kilocycles with daytime power of 50,000 watts oriented toward the Pacific Ocean. During this time, there were about 95 disc jockeys, 17 program directors, six general managers, three talk show hosts, and two ownerships. Here is their story...

1955-1965 – THE HITPARADERS

1954

A Special note:

Ernie Meyers (6-9 AM) had a goal in life to become a horse jockey, but he lacked the small stature to be successful and became a disc jockey instead. He got into radio as  an announcer on Armed Forces Radio while serving in the Army during the Korean War. He arrived in San Diego in 1954 and became the morning man at KCBQ. Disliking the Bartell format when they bought the station in 1955, Ernie left and was hired at XEAK in 1958, commuting from his home in Imperial Beach to the transmitter in Mexico. Later, he would be at KOGO radio for 19 years and KSDO for 14 years. He concluded his career at KPOP in November 2000 after suffering a stroke.

1955

KCBQ ranked seventh as a CBS affiliate in the San Diego market when the Bartell Group bought the station in 1955. The Bartell Group dropped the affiliation and introduced Top 40 radio into the San Diego market in December 1955. The Bartells took an active role in the daily operations of the station. Their management-driven approach created consistency and success, with few interruptions, over the decades they owned the station.

The charter Top 40 disc jockeys in December 1955 under Program Director Allen Heacock were:

Ralph James (Ralph Torres) (6-9AM) the original morning man remained with the station until May 1960.

Ralph eventually moved on to Hollywood where he did voice-overs. Among his credits was the voice of Orson on the show Mork and Mindy.

Jim O'Leary (9-12 and 7-9PM) had been with KCBQ during the CBS affiliate days of the early 1950s. He had left for KBIG in Avalon, but returned at the request of the Bartells in 1955. His employment typified the multiple shift requirements employed by radio at that time.

Harry Martin (12-4PM) began his radio career at KGBC in Galveston TX during the late 1940s. He  was recruited by the management of KLAC in Los Angeles to come to the west coast and was going strong when he was drafted into the Army in 1950. He was hired at KCBQ in 1955 and was dubbed Happy Hare by colleague Don Howard. He held this shift until February 1958 when he took over the 6-9 AM shift. He was transferred to WADO in New York City by Bartell in October 1959 and returned in July 1960.

Don Howard (Max Schwimley) (4-7PM and 9-12PM) was already well known in San Diego. He started at KSDO on September 17, 1948 before being hired by KCBQ on November 1, 1955 for the princely sum of $25,000 per year. He continued doing  an early evening in-studio shift then a live shift from a nightclub he owned at 59th and El Cajon called Don Howard's Tempo. In 1956, he moved to the mid-morning shift that he held until he left the station in August 1963. Don has been characterized as a 'classy' person. He had theme songs at the beginning and end of his shows. He was the mentor of many local boys who became prominent radio personalities. Don loved jazz and equally hated the rock n roll music format. Don went to KOGO AM 600. He died in 2001.

In its first year, the station had no overnight person to do the graveyard shift. Instead, a taped Hit Parade was broadcast.

Jack Vincent (John Vincent Oatsdean) began his career at KXO in El Centro, CA. He was hired as the station engineer by Ralph Menard in 1955 when the station was still under the Salek ownership. Jack had been at KFXM in San Bernardino. It was the intent of station management to have him do the overnight shift. However, technical staff were in a different union than air talent staff and a conflict was deemed to exist that required extended negotiation.

1956

The station began publishing a weekly *Hit Paraders* survey in 1956 that featured a top 20 play list on a single-sided 8 1/2 X 11 sheet of paper. The station call letters appeared in a rectangular box with an elongated 'b' that became known as the Bartell logo.

Jack Vincent (12-6AM) was the only lineup change for the year when he finally gained clearance to do air work. Because of his multiple abilities, he earned a salary of $135 per week. This was second only to Don Howard, earning $225 per week, from being the highest paid radio personalities in the San Diego market. Jack did his shift at the transmitter in Santee, while all others did their shows from the studio at 7th and Ash. He was always a gentleman. But he had a pet peeve about DJ's in the next shift being late after he had been working six hours on the air.

It became well known that Jack enjoyed visitors at the transmitter to break the boredom. Fans like Tommy Irwin and Phil Boles would make visits whenever they could find someone to drive them there. This Clark Gable look-alike used a standard phrase "as we bounce right back on KCBQ" after breaks, remained in that shift through 1968. But he wasn't done. He served another 15 years after that as the station engineer. In 2003, this venerable 85 year-old still lives in Santee. He made his last radio appearance in 1975, and appeared at the 1990 DJ reunion. He can still be found on Saturday nights enjoying a cigar and beer while playing pool at the home of Tommy Irwin (Shotgun Tom Kelly).

1957

The lineup was realigned in January for the first time since the station began Top 40.

Ralph James (6-10 AM) continued as the morning man.

Don Howard (10 AM-Noon and 4-7 PM) continued to do split shifts.

Harry Martin (Noon-4 PM)

Jim O'Leary (7-12 PM) He departed the station soonest in February 1957 after being homesick for Avalon and KBIG. He later went to KMPC in Los Angeles and died suddenly in 1963.

Jack Vincent (12-6 AM)

The lineup as of September 1957 featured numerous split shifts and the debut of Ralph James in a night shift.

Harry Martin (6-9 AM and Noon-2 PM)

Don Howard (9-Noon and 4-6 PM)

Ralph James (8-12 PM)

Jack Vincent (12-6 AM)

Earl McRoberts (2-4 PM and 6-8 PM) arrived in February 1957 to work a split shift to replace Jim O'Leary. He went to KDAY in Los Angeles in February 1958, and was later a long time news anchor at KFWB.

1958

The daytime power was increased to 50,000 watts in 1958. The night time power remained at 5000 watts. Bartell could have been assigned 10,000 day and 10,000 night watts that would have allowed antennas to be directed to populated areas. As it was, much of those 50,000 watts were directed westerly out to sea in order to meet FCC requirements. The station still provided the strongest signal emanating from the San Diego market, and provided Lee Bartell the prestige of a 50,000 watt station. KCBQ launched a contest of hiding $50,000 for a listener to find to promote the power increase. The check was hidden in a lipstick container under a specific telephone pole at the end of Swift Ave. Clues were given over the air about the location of the cash. The *Treasure Hunt* contest was copied from radio station KLIF Dallas, TX (only they used a soda bottle instead of a lipstick container), and met with amazing success in December 1956. This contest had a deadline of November 10, but ended on Halloween when a listener found the check and validated it at the radio station. The contest had the greatest audience participation in San Diego history up to that time.

The station expanded its Hit Parader Survey to feature the top 40 play list rechristened as the *KCBQ Hit Parade Survey* that began with Issue No. 1 on February 23, 1958. Black and brown inks were used on an 8 1/2 X 11 sheet that beneath the play list featured peel off labels for listeners to apply to their records.

Soon, the survey included a list of record picks, selected by Music Director Phyllis Sandin, under the name of each DJ. The DJs had no input on the choices. The Bartell logo was inset within the shadow of a guitar. The station continued to develop programming and promotions under a stable staff.

The lineup as of February 1958 was as follows:

Harry Martin (6-9 AM and Noon-2 PM)

Don Howard (9-Noon and 4-6 PM)

Ralph James (2-4 PM and 6-8 PM) assumed the shifts of Earl McRoberts.

Lucky Lane (Dick Boynton) (8-12PM) came for a brief stay in February to assume Ralph James' shift.

Jack Vincent (12-6 AM)

Scotty Day (Jim Neill) 'Scotty Day' was a Bartell name used in several cities. This Scotty arrived in April to begin the first of three phases in his career at KCBQ. At this time, Scotty did production work and weekend air shifts. His smooth style has made him the all-time favorite DJ of many listeners. Scotty was not flamboyant but could show a temper. He often mispronounced the names of music groups, but no one cared. He opted to be paid bonuses with Bartell stock rather than cash and loved to  play the stock market. He was a wealthy man by the time he left the station in 1970. Besides being a production manager, music director, and program manager, he wrote many of the station jingles and sang in many of them. He did the weekly *hitparader* countdown each weekend.

The lineup of September 1958 involved the following adjustments:

Harry Martin (6-9 AM and Noon-2PM)

Don Howard (9-Noon and 4-6 PM)

Ralph James (2-4 PM and 6-8 PM)

Jerry Walker (Harry Walker Birrell) (8-12PM) arrived from WNOE New Orleans in September to  replace Lucky Lane and would remain until 1965 when he went to WINW in Canton, OH. His air name arose to avoid confusion with Harry Martin; and, because management didn't care for his last name, they opted for his middle name. After joining KCBQ Jerry made a wise-quack on the air that brought PD Al Heacock running in asking "What was that?" After an instant audition, Heacock told him to keep it up. A contest was held to name the duck and the winning entry was Casey B. Quack. Casey was given full DJ duties including song selections in the weekly survey

Jack Vincent (12-6 AM)

Scotty Day (Weekends and production)

The staff was highly productive in other ways too. Ralph James had a daughter born on October 13, Jerry Walker had a son born on October 14, and Don Howard had a son on October 15.

1959

Jonathon Kirby replaced controversial Ben Shirley as New Director. Kirby produced a landmark news series called *The Critical Years* when a 17 year-old named 'Tommy' appeared at the station to provide insight about juvenile delinquency, leading to gang activities, from the perspective of a gang member.

The series was broadcast over the seven-station Bartell network and won recognition by the Governor for its value as a public service.

By October, the Bartell Group sought to establish a station in New York City. Program Director Alan Heacock and Harry Martin were transferred to format WADO in the Bartell image. Because the station had a weak transmitting signal in a market where strong Top 40 stations were already established, the project was abandoned after a few months.

This enabled Scotty Day to assume the position of Program Director of KCBQ, as the second phase of his long career with the station. He always preferred air and production work over management. However, Lee Bartell handled all personnel decisions and Phyllis Sandin handled music selection.

The lineup had no changes until November:

Ralph James (6-9AM and Noon-2 PM) became locally known as 'Big J' during this time because due to his emerging portly build.

Don Howard (9-Noon and 2-4 PM)

Jerry Walker (4-8 PM)

Johnny Holiday (Ed Phillips) (8-12PM) arrived from WYDE Birmingham, AL in October 1959 to replace Harry Martin. 'Johnny Holiday' was an often-used Bartell name. This Johnny was the fourth DJ in this slot in four years and remained there until 1962. He was a man of many distinguished names. He used his real name in Birmingham, Johnny Mitchell at KHJ in Los

KCBQ RADIO AM 1170

 Angeles where he enjoyed enormous success, and Sebastian Stone at KFRC in San Francisco. He would have the highest ratings in the San Diego market during his tenure at KCBQ and remained with the station until 1965. He then joined KHJ. He died on November 11, 1987 of a heart attack.

Jack Vincent (12-6 AM)

Scotty Day (Weekends)

1960

The weekly survey was modified as the *KCBQ Hit Parade* Top 40 on a 4 1/2 X 7 two-sided format. Photos of all the DJ staff were on front. Survey numbers were continued from the previous design in sequential order.

Changes were made prior to May that were mainly shift realignments:

Jerry Walker (6-9 AM and 2-4 PM)

Don Howard (9-Noon and 4-6 PM)

Ralph James (Noon-2 PM and 6-8 PM)

Johnny Holiday (8-12 PM)

Jack Vincent (12-6 AM)

Scotty Day (Weekends)

The lineup in May had a new addition:

Jerry Walker (6-9 AM and Noon-2 PM)

Don Howard (9-Noon and 2-4 PM)

Shadoe Jackson (Jerry Swearingen) (4-8PM) arrived in May from KDEO when Ralph James became news director. The name 'Shadoe' was used across the country by DJs working during the early evening shifts, most notably by Shadoe Stevens at KRLA in the early 1970's. This Shadoe would later lock down the 8-12PM shift from 1962 until June 1965. During that time he promoted the submarine races at the San Diego River and was ordered by Lee Bartell to stop because teens were assembling at the river.

Johnny Holiday (8-12 PM)

Jack Vincent (12-6 AM)

Scotty Day (Weekends and production)

The lineup in July had a returning DJ:

Harry Martin (6-9 AM and Noon-2 PM) returned in July from WADO New York City. This was his second tour with the station. This was the last split shift employed by the station.

Don Howard (9-Noon)

Jerry Walker (2-5 PM)

Shadoe Jackson (5-8 PM)

Johnny Holiday (8-12 PM)

Jack Vincent (12-6 AM)

Scotty Day (Weekends and production).

1961

The lineup established in January remained intact until June 1962. The station was being pushed in the ratings by KDEO throughout 1961, and appears to have relinquished the market lead at times during the year.

Harry Martin (6-9 AM)

Don Howard (9-Noon)

Jerry Walker (Noon-4)

Johnny Holiday (4-8 PM)

Shadoe Jackson (8-12 PM)

Jack Vincent (12-6 AM)

Scotty Day (Weekends and production)

1962

The weekly survey dropped the Bartell logo in

favor of a stylized *The Big KCBQ Survey*. The overall format and numbering sequence remained the same. The station resumed its dominance in local ratings.

The lineup did not change until June as follows:

Jerry Walker (6-9 AM)

Don Howard (9-Noon)

Johnny Holiday (Noon-4 PM)

Seamus Patrick O'Hara (4-8PM) 'Friend of the Leprechan' arrived from KYA San Francisco in June when Harry Martin left for Cleveland. He would eventually become the morning man, when Seamus was then dubbed King Seamus by station owner Lee Bartell. This was in keeping with 'titles' bestowed upon, or assumed by, other radio personalities throughout the nation at the time.

Shadoe Jackson (8-12 PM)

Jack Vincent (12-6 AM)

Scotty Day (Weekends and production)
In the category of *what were they thinking*, the station gave away a burro at Christmas along with $117.00 in cash to buy hay.

1963

There were no lineup changes until August when Don Howard left:

Jerry Walker (6-9 AM)

'Dollar' Bill Bishop (Gerald Blume) arrived in August from WDRC Hartford, CT to replace Don Howard in the 9-12 noon shift when Don went to KOGO. He started his radio career during high school. All other staff remained. He also remained until June 1965. By 1969, Bill was on KFMB radio as Jerry Bishop. He currently does voice-over work that includes the Judge Judy show.

Johnny Holiday (Noon-4 PM)

Seamus Patrick O'Hara (4-8PM)

Shadoe Jackson (8-12 PM)

Jack Vincent (12-6 AM)

Phil Roberts (Weekends) would later resurface as an announcer at Sea World.

Scotty Day (Production)

1964

Artists began appearing on the cover of the weekly survey when Beatlemania erupted in March.

The only changes in the lineup involved realigned shifts in December:

'King' Seamus Patrick O'Hara (6-9 AM)

Bill Bishop (9-Noon)

Jerry Walker (Noon-4 PM)

Johnny Holiday (4-8 PM)

Shadoe Jackson (8-12 PM)

Jack Vincent (12-6 AM)

Scotty Day (Weekends)

1965–1967 – BLOOD IS THICKER THAN WATER

The years 1955, 1965, and 1971 mark pivotal milestones for KCBQ. The significance of 1955 is obvious for the inaugural of the Top 40 format. 1971 will be discussed later. In 1965, several paths were transecting that would redirect the course of the station.

First, Bill Drake's *Boss Radio* format caught on as the most popular concept in radio. He was able to hasten the tempo of the hourly program clock in a way to air more music. All accounts suggest this

KCBQ RADIO AM 1170

was reflected in staggering audience ratings that spooked Lee Bartell. He took stock of his *Hitparaders* that dated back to the 1950s, and decided it was time to move his product in a new direction. Gary Allyn wrote, "Mr. Bartell believed in the 'Family Radio' concept which was somewhat dull in presentation for the times . . . KCBQ would garner more adults, while KGB would win more teens." Significant changes within the air staff occurred on almost a monthly basis. Some of the staff were openly encouraged to find employment elsewhere. Others went on vacation and returned to find someone else had their job.

Meanwhile, changes were occurring among the professional and personal lives of the staff that would have mandated change anyway. Lee Bartell had been functioning as the station General Manager. When the Bartell Media finalized acquisition of MacFadden publications, older brother David Bartell insisted that Lee, over his objection, move to New York to become president of Bartell Media, with David as chairman of the board. Rather than hire an experienced manager to run the station, Lee put the operations in the hands of his Music Director, Phyllis Sandin. This would enable Lee to use the station, and his family who remained in San Diego, as an excuse to return rather often. Ms Sandin had no training or skills in radio management. As KGB captured the market, KCBQ was adrift without adequate day to day management.

The authority of the station Program Director continued to be limited. There was a rotation of DJs who acted as program managers that extended through 1967. Hiring and firing decisions were still being handled by Lee Bartell from New York that often resulted in a DJ showing up for a shift without anyone knowing anything about them.

The real power of the station was vested on a daily basis in the person of Music Director Phyllis Sandin. In this role, she asserted veto power over music selections. She guarded the studio entry to prohibit grade school boys, such as Tommy Irwin and Phil Boles, from visiting their broadcast idols. She inspected all incoming mail and rummaged through staff desk drawers. Its no wonder she did not have enough hours in the day to accomplish all her self-assigned duties. Johnny Holiday hated her and advised all incoming staff to be on their guard around her. Gary Allyn admonished Bartell that if he trusted him enough to go on the air under his (Bartell's) FCC license, he needed to trust his judgement that Phyllis had to go. Lee and Phyllis were close and had a devotion to each other that extended far beyond the workplace, according to staff who were there at the time. She stayed.

KFWB arranged a Beatles concert at Balboa Stadium in August and coordinated the event with KCBQ and KGB for promotions. KDEO got in on the fun by having a band play on a flat bed truck as it drove through various communities to promote the event. Using his connections, KCBQ DJ Lord Tim Hudson managed to make an appearance on the stage, but the loud audience made it clear they were there to see and hear the Beatles, not local or LA radio personalities.

The survey format and numbering sequence remained the same, but now featured a single DJ or artist each week because there would be many new staff to introduce to the San Diego listening public. The lineup through May 1965 featured the *Hitparaders* staff who had been with the station for years. The staff was now going to be 'boxed' around.

ROUND ONE – *May*

The first quarter remained stable. The *Hitparaders* provided familiarity and stability for the listeners. Despite their standing with listeners, and despite their loyalty to the ownership, they were about to be phased out.

Johnny Holiday (6-9 AM) replaced Seamus O'Hara in this shift.

Bill Bishop (9-Noon) would move on to later become the voice of Disney.

Jerry Walker (Noon-4 PM)

Seamus Patrick O'Hara (4-8 PM) was losing ground to a severe drinking problem. It was arranged for him to get treatment and he left the air on May 23.

Johnny Williams (4-8 PM) was the first of the 'new blood' to arrive. He was involved in a contract dispute between rival stations KIMN and KBTR in Denver. The owner of  KIMN, Ken Palmer, set John up with a job at KCBQ and picked up his moving costs. The plan was for Williams to come in as program director and do the morning shift, as Johnny Holiday was closing his deal with KHJ in Los Angeles. But with the O'Hara problem at hand, Holiday moved to the morning shift and Williams replaced O'Hara in the afternoon shift. Williams wanted to make changes to the method of song selection and immediately got into a 'spitting match' with Phyllis Sandin over what music would be played. He took his complaints to Lee Bartell who said "you're not going to work out here..." and promptly fired him after three weeks of service. After doing a couple of weekends at KRLA, Williams was hired by Bill Drake and was soon joined by Johnny Holiday as "Johnny Mitchell" at KHJ. He now masters the Satisfaction 440 registry website that charts the career paths and whereabouts of Top 40 disc jockeys all over the country.

Shadoe Jackson (8-12 PM) may have worked a 4-Midnight shift when the station was short of staff in May. He was on his way out as part of the purge of the 50s *Hitparaders*. He would resurface at KOGO by 1970 as the program director Jerry Jackson and hire his former colleague Scotty Day

Jack Vincent (12-6 AM)

ROUND TWO – *June*

The station needed a theme to counter *Boss Radio* format so they adopted the extremely successful *Good Guys* theme patterned after WMCA in New York. An overhaul of the air staff was well under way as the '*Hitparaders*' were phased out.

Tom Murphy (6-9 AM) arrived in mid-June to take this shift when Johnny Holiday took the 3-7 shift. He and Gary Allyn were roommates when they arrived in town. Tom was older than the other staff. He performed on the original Murphy and Harrigan show in the 1950s.

Stan Richards (9-Noon) arrived from KONO San Antonio in mid-June to replace Bill Bishop in the mid-morning shift. By this time, he was highly respected in the industry for having a good programming mind and doing great contests. It was Stan that arranged jobs for Gary Allyn and Tom  Murphy. But by this time, he was suffering from family and financial stress that affected his health. He began feeling dizzy and having fainting spells and was eventually hospitalized with depression. Lee Bartell fired him while he was in the hospital. He left the business and moved in with his parents in Memphis, TN for a year. Later, he resumed his career and became a successful sports announcer and voice-over artist. He died of heart problems in the 1990s.

Gary Allyn (Gary Allyn Hempstead) (Noon-3PM) began his career in radio at age 17. He majored in various media fields in college, then got his first practice at radio in the Army. His achievements had already included booking the first US appearance of the Rolling Stones in 1964. He arrived  from KIMN Denver in mid-June to assume part of Shadoe Jackson's shift and would be a name highly associated with the station through 1970. At in a height of 5'7", he proclaimed himself as the World's Tallest Midget. Owner Lee Bartell, being at 5'4", found the term offensive so Gary became Leader of the Little People instead.

Johnny Holiday (3-7 PM)

Jerry Walker (7-Midnight) left at this time and took Casey B. Quack along with him, never to be heard on the air again. After a brief period selling

KCBQ RADIO AM 1170

life insurance, Jerry went to WINW in Canton, OH in 1966, then later joined KNX Los Angeles in 1968 as a news anchor where he remained for 30 years.

Jack Vincent (12-6 AM)

ROUND THREE – *July*

The station even made a direct link to the British invasion sweeping the nation.

Tom Murphy (6-9 AM)

Stan Richards (9-Noon)

Gary Allyn (Noon-3 PM)

Lord Tim Hudson (3-7PM) arrived in July to assume a new shift. Bartell had hired Hudson and pushed a major promotion campaign touting Hudson as the first British DJ in the region and the embodiment of the British invasion into rock n roll. The station housed him at the El Cortez Hotel.  Hudson was in his mid twenties and presented quite a different look around the station. He wore a 'Beatle-style' haircut and bell bottom pants. He was quite a charmer with girls, and was quite adept as a fast-talking self-promoter who boasted of his inside connections with The Moody Blues. But contrary to his claims, he had never been on radio before and had no experience. He could not work the controls, start records, or punch up ads. All he could do was sit there and talk. His arrival at KCBQ represented one of the largest promotional and staffing disasters. At a concert hosted by KGB radio, he was recognized to the audience by fellow Brit Ian Whitcomb, who was unaware that he was from a competing station. Whitcomb was blacklisted from further KGB events.

Johnny Holiday (7-Midnight) received an opportunity to become a Boss Jock at KHJ in Los Angeles and took it in mid July.

Jack Vincent (12-6 AM)

ROUND FOUR – *August*

Tom Murphy (6-9AM)

Stan Richards (9-Noon)

Gary Allyn (Noon-3PM)

Lord Tim Hudson (3-7 PM) left by October to KFWB Los Angeles. He now works as a stand-in for actors in Hollywood.

'Gentleman' Jim Mitchell (7-Midnight) arrived from KMEN San Bernardino to replace Johnny Holiday. He had also worked at KGB for three months in 1963. The pedigree of 'Gentleman Jim' began with 'Gentleman' Jim Markham from KMEN, which Jim Mitchell inherited.

Jack Vincent (1-6 AM)

Seamus Patrick O'Hara (Weekends) returned to the weekends after rehab and the station put him up at the El Cortez Hotel. But he fell off the wagon and was just too drunk to work anymore. The station released him and he ended up at Patton State Hospital. It is not known how his life concluded.

ROUND FIVE – *September - October*

Jack Hayes (Jack Hayes) (5:30-9AM) Dubbed "The King" by Lee Bartell (after Seamus O'Hara left) arrived from KFWB Los Angeles in September to fill the shift vacated by Tom Murphy. Lee Bartell had promised Hayes a series of raises if he  improved the ratings when he got to KCBQ. The first two monthly ratings showed significant growth and Hayes sought the money he was promised. Bartell responded with "Nobody could have that immediate an effect on ratings, we'll wait six months." Hayes replied, "That's not what you promised, pay me or I'm gone."

Scotty Day (9-12AM) moved from his program director duties to the broadcast booth, replacing Stan Richards in October, where he would remain until 1970. This was the third phase of his career at KCBQ. He was typical of mid-day jocks who were just themselves rather than taking on a persona.

'Gentleman' Jim Mitchell (12-3PM)

Tom Murphy (3-6PM) replaced Lord Tim Hudson in this shift.

Gary Allyn (6-9PM) continued in this shift in October, even after becoming PD.

Johnny Solo (9PM-1AM) arrived in September and held this shift during the fall. He was one of several Johnny Solos in radio at the time. The name originally took root from the name of a character in Ian Fleming's book and movie *Goldfinger*, and was heavily popularized by Napoleon Solo character in *The Man from UNCLE* TV series. He was not the Johnny Solo Gary Allyn worked with at KONO. He was another Bartell hire that came, stayed briefly, then left without a trace. His real name is not known.

Jack Vincent (1-5:30AM) maintained the overnight shift. Jack shunned the downtown studio at 7th and Ash, and never attended staff parties. Gary Allyn decided to visit the transmitter in Santee one night to see if Jack even existed. After meeting Jack, Gary was amazed to find that every 45 record, brought out to the transmitter each week for airplay since 1956, was stacked neatly against the wall on broom handles.

1966

KCBQ was floundering to the point that advertising sales had dropped considerably since Lee Bartell relocated to New York. This was a great concern to the Bartell Media board. David Bartell told Lee to either hire a general manager or the board would sell the station. Using contacts through Blair National Representatives, the leading radio sales organization at the time, Lee met and hired Stan Torgerson as the new General Manager. Stan was hired based on his sales skills rather than any programming ability. He focused on finance decisions and left programming to the Program Director.

There were other changes too. Herb Mendelson had been appointed as president of the radio division for Bartell Media. He would visit to discuss sales and programming. Former KFWB DJ Ted Randal was a program consultant for the stations in small markets across the country. He took on KCBQ this year. Randal's connections enabled access to new record releases ahead of most competitors. When not consulting, he maintained a law practice in Los Angeles.

Meanwhile, Phyllis Sandin's job position was changing. She was introduced to the new GM as Lee Bartell's 'secretary', and that is how she was considered. Her bitterness grew as she lost all decision-making power over music, personnel, and management as her duties were relegated to menial tasks, such as handling telephone installations. Her relations with others became cold.

The weekly *Big Q Survey* became a single sheet flyer that featured an ad or a DJ on the back. The Bartell logo replaced *The Big Q* heading on surveys in the summer. The size and numbering sequence (now nearing Issue 500) continued. The station also streamlined their jingle package.

Jack Hayes (5:30-9AM) quit in February, (but soon returned), over money issues with Lee Bartell and went into program consulting.

Chuck 'Huckleberry' Clemans (5:30-9 AM) arrived in February to replace Jack Hayes. Huckleberry had developed his career at KMEN San Bernardino.

Scotty Day (9-12AM)

Jim Mitchell (12-3PM)

Tom Murphy (3-6PM) left in August. He resurfaced in January 1967 at KBLA in Los Angeles as Bobby St. Thomas.

Gary Allyn (6-9PM)

KCBQ RADIO AM 1170

Robin Scott (9-Midnight) arrived from KDEO in mid-February to replace Johnny Solo. Emphasis was placed on his first name to capitalize on the *Batman* craze of 1966. He later went to country stations KBBQ and KLAC as Bob Jackson. Robin was considered a good-natured man and an adequate DJ. He overcame physical constraints caused by polio to enjoy a successful radio career. He now lives in Wichita Falls, TX.

Jack Vincent (Midnight-5:30AM)

Jeff Crane (Scotty Palmer) (Weekends) hung around the studio as a teenager, then transitioned into the news department, then to a DJ in May.

The shifts were realigned in August as Jack Hayes returned as Program Director.

Chuck 'Huckleberry' Clemans (5:30-9 AM)

Scotty Day (9-Noon)

Jim Mitchell (Noon-3PM)

Jack Hayes (3-6 PM) replaced Tom Murphy and served as program director. The accomplishments of KCBQ have been celebrated in different forms over the years, none being more prominent that the 1990 Reunion. The photo at left features Jack Hayes and Lee Bartell. So many personalities have died since that reunion, including Lee Bartell, who died a few months after this photo was taken. The Reunion left a lasting contribution to the historic record of the station as many personalities used this last chance to retell their story on air and on video tape.

Gary Allyn (6-9 PM) left in mid August. His shift was eliminated for a short time.

Robin Scott (9-Midnight)

Jack Vincent (Midnight-5:30AM)

Jeff Crane (Weekends)

September brought further changes as a talk show format was inaugurated:

Chuck 'Huckleberry' Clemans (5:30-9 AM)

Scotty Day (9-Noon)

Jim Mitchell (Noon-3PM) went to KHJ in Los Angeles in October under the name of Jim Lawrence, (because they already had a Johnny Mitchell on staff), to do news. He went on to a successful television news career and now teaches journalism law at the University of Arizona at Tucson.

Jack Hayes (3-6 PM) gained a coveted position that had not changed since the Top 40 format began at KCBQ. Jack was named Music Director, succeeding Phyllis Sandin, with complete control over the music selection on the play list.

Robin Scott (6-9 PM)

Newsman ***Pat Michaels*** (9-Midnight) had been with the station since 1964. Pat had done a half-hour documentary about police brutality which won a Golden Mike award for KCBQ. When the 9:00-Midnight ratings bottomed out, Lee Bartell decided to break format and start a talk show around Christmas 1965. Michaels was busy already doing the morning news at KCBQ and the evening news at XETV-6 based in Tijuana. The idea was to have a mix of records and 'pithy' comments from Pat, but the *I Hate Pat Michaels* program took off and the number of callers left no time for records. Lee Bartell, who was very liberal, directed Michaels to espouse those views on the air, hence the name of the show in conservative San Diego. This marked one of the very earliest forays into the talk radio formats now common today.

Bumper stickers emerged around town promoting the show. Just before the gubernatorial election of 1966, Michaels engaged in his usual combative discussion with a listener reporting that Pat Brown was speaking with elementary students and had equated Ronald Reagan with John Wilkes Booth by reminding them that an actor killed Abraham Lincoln. This report was accurate and got a lot of publicity in the waning weeks of the election, which Brown ultimately lost. And we heard it first on the Pat Michaels show.

Jack Vincent (Midnight-5:30 AM)

Jeff Crane (Weekends)

The following changes occurred during November:

Chuck 'Huckleberry' Clemans (5:30-9 AM)

Scotty Day (9-Noon)

Happy Hare (Noon-3) returned to the station on November 21 to replace Jim Mitchell. Hare disputes this information, but the recollections of several listeners, including Bill Earl who kept a diary of station events, confirm Hare's return.

Jack Hayes (3-6 PM)

Barry Boyd (6-9 PM) arrived from KFXM in San Bernardino took this shift when Robin Scott left. Robin resurfaced as Bob Jackson at KBBQ in Los Angeles in June 1967.

Pat Michaels (9-Midnight)

Jack Vincent (Midnight-5:30 AM)

Jeff Crane (Weekends)

1967 – Time for *Action*

Bartell Media was choking the station on minimal operating budgets that resulted in three program directors quitting during the year. Hooper ratings for June 1967 showed a 4 share in the afternoon drive, while Bob Elliott of KGB enjoyed a 24 share. The station was ranked 12th in a six station market as Los Angeles stations grabbed local market share.

The station started branching into different formats and promotions in an effort to restore declining ratings. The austere *Hit Parade* Top 40 card ceased production in April.

The new '*Action*' theme began in April as the DJs were referred to as *Action* Men and the survey became the Action 30. 1967 saw sweeping changes as the 'societal counter-culture' emerged. DJ Bobby Wayne led an *Action* Fun Trip to San Francisco's Haight-Ashbury District with a contest winner and were pictured on the streets of San Francisco with real hippies. On the local scene, the station promoted the Beach Patrol that made rounds to area shore lines giving away prizes. The station also produced Scene Magazine that provided a behind-the-scenes look at station operations, though it probably didn't cover the escapades of the Fun Lovin' crew at their favorite haunts; the Brigantine and the Chart House on Shelter Island.

The staff and management changed often in 1967. The lineup emerged during January were:

Chuck 'Huckleberry' Clemans (5:30-9 AM)

Scotty Day (9-Noon)

Barry Boyd (Noon-3 PM) moved to this shift upon the arrival of Larry Mitchell.

Jack Hayes (3-6 PM)

Larry Mitchell (Frank Lawrence Coviello) (6-9 PM) was hired in January by Jack Hayes when Harry Martin went to Detroit. Larry had worked with Hayes at KLIV in San Jose in 1960. They became not only good friends, but also business partners in concert promotion and program consulting.

Pat Michaels (9-Midnight) *I Hate Pat Michaels* talk show

Jack Vincent (12-6 AM)

The following adjustments occurred in April:

Chuck 'Huckleberry' Clemans (5:30 –9 AM) left in April to pursue a career as a stockbroker. He resumed his career at KMEN in 1974 and was soon hired by Ron Jacobs at the 6-9 AM DJ at progressive rock station KGB. While there, he developed the Tyrone the Frog character.

Scotty Day (9-Noon)

KCBQ RADIO AM 1170

Barry Boyd (Noon-3 PM)

William F. Williams (3-6 PM) arrived from KMEN in San Bernardino by April 13 to replace Jack Hayes as PD and assume this shift. After a brief stay, he went to KBLA to replace Bobby St. Thomas (aka Tom Murphy) when Tom went to KFWB. Williams stayed at KBLA after the change to a Country format as Bill Williams. Jack Hayes was back as PD and in this shift by April 19.

Jack Hayes (3-6 PM) began and ended the month in this shift and as PD.

Larry Mitchell (6-9 PM)

Pat Michaels (9-Midnight) the liberal-orientation of the *I Hate Pat Michaels* talk show had become an irritation to conservative-minded GM Stan Torgerson so he fired Michaels, not knowing that the program concept was at Lee Bartell's direction. Bartell apologetically supported his GM's decision and provided Michaels with an uncharacteristically generous six months severance pay. Pat Michaels was soon recruited to KGO in San Francisco to fire up his listeners with conservative positions. The pragmatic Michaels continued to gain lucrative jobs over the years when, in 1984, he was approached about participating in an upstart talk radio network. The money and location weren't right at the time for Michaels, but he recommended a guy in Sacramento whom they listened to and hired. His name . . . Rush Limbaugh.

Jack Vincent (12-6 AM)

Jeff Crane (Weekends)

The following adjustments occurred in May as follows:

Barry Boyd (6-9 AM) took this shift when Chuck Clemens left.

Scotty Day (9-Noon)

Larry Mitchell (Noon-3 PM) served briefly as PD after Jack Hayes left.

Jack Hayes (3-7 PM) Once again, compensation was based on ratings, there was slight improvement, and the compensation failed to follow, so Jack quit again. In September, Bartell asked Jack to return. This time he refused, but did enter into a consulting agreement until he went to work for KNBR San Francisco. He spent several years with KNBR and remained close friends with Lee Bartell despite money issues. Jack still does radio programming consulting. He has a home in La Jolla with an office and motor yacht in Marina Del Rey, CA

Bobby Wayne (Wayne Satterfield) (7-11PM) arrived from Minneapolis in May, but his career was centered mainly at stations in Cincinnati. Bobby was a creative talent who played the Top 40 hit list during the early hours of  his show, then developed a show called 'Funderground' that featured more obscure artists like Velvet Underground, Spirit, Chambers Brothers, Love, and others that extended to midnight. He was an entertainer, rather than a song announcer. He created characters, such as Hound Dog that sounded very similar to Wolf Man Jack. He used 'Chicken Man' the crime fighter created by Dick Orcan of WCFL Chicago to add dimension to his broadcast. He promoted his show as the 'Wayne Train raising Cain'. His show was aimed at high school students that made him a local teen idol. He gave time for 'Q-spondents' to call in and report on their high school activities, and participated in the station 'Beach Patrol' to promote KCBQ with teens at area beaches. The beginning of the turn-around for KCBQ is credited to Bobby Wayne. The photo above features Bobby Wayne with Melody Galore. His future was bright, but he also had a drinking problem. He would often be busy on the phone and not paying attention as a record would run out, or carelessly drop a needle on the grooves. This led to his firing by Program Director Mike Scott in June 1968. He was at WCBS-FM from 1969 to 1974 and was ultimately fired for too many foul ups and "bad days". He died at his Florida home in Sept. 1990.

Jack Vincent (11-5 AM)

Jeff Crane (Weekends)

The following adjustments were occurring in June as Jim O'Brien became PD:

Barry Boyd (5-9 AM)

Scotty Day (9-Noon). Scotty worked the Noon-3 PM shift between the departure of Chuck Clemans and the arrival of Jim O'Brien.

Jim O'Brien (Jim Oldman) (Noon-3 PM) arrived from KLIF in Dallas at the same salary. Jim was said to be very personable with a smile in his voice. He emphasized the 'Q' as he promoted the station. For example: "Its 8 O' Clock in Q Country with more non-stop music". He was better known at WFIL, Philadelphia where he was from 1970 to 1983 on radio and TV. He followed Ron Jacobs as the PD of KHJ Los Angeles in 1969. Jim was killed in a sky diving accident in 1983. His legend lives on through his daughter Peri Gilpin who plays Roz Doyle on the Frazier TV series.

Mike Scott (Private) (3-7 PM) got his first radio job at KTSA San Antonio as Mike Bell the 'bellboy' after his discharge from the Air Force. After a stint at KONO, PD Stan Richards (later of KCBQ) hired Mike at KLIF in Dallas, TX in September 1966. Gary Allyn  was doing weekends there at the time. Mike was shipped to Libya and a Frank Bell was hired at the station. Upon his return, his name had to be changed to avoid confusion and Mike 'Scott' emerged. He became the station music director. Jim O'Brien had worked with Mike at KLIF and hired him at KCBQ in July at the same salary. He quickly gained the nickname 'Big' Mike Scott because at 5' 11" he towered over most of the staff. Mike notes that upon his arrival, "the station lacked any soul. It was a machine; a business where the only meaningful bottom line was dollars." He was a key resource for KCBQ to gain national prominence. He did afternoon and weekend shifts until he became PD in June 1968.

Bobby Wayne (7-11 PM)

Jack Vincent (11-5 AM)

Jimmy Mack (Jim Talley) (Weekends) arrived in July for his first tour at KCBQ.

A KCBQ edition of Robin Leach's national GO Magazine began in July that promoted local station identity and features, including the weekly survey. The play list was reduced from top 40 to Top 30.

By September, the line-up featured the following staff under the direction of new Program Director Stu Langer. Stu had an extensive writing background, but he was ill suited for radio. He had no sense for broadcast media and would often use cuts from comedy albums as filler through the day, throwing the broadcast tempo off. Meanwhile, KGB was humming like a well-tuned machine.

Barry Boyd (5-9 AM)

Scotty Day (9-Noon)

Jim O'Brien (Noon-3 PM) had a falling out with the Bartells over operating budgets and left in September.

Mike Scott (3-7 PM)

Bobby Wayne (7-11PM)

Larry Mitchell (11 PM-2AM) did news as well as host this shift

Jack Vincent (2-5 AM) continued to anchor the overnight shift.

Jimmy Mack (Weekends)

The lineup by the end of the year was as follows:

Barry Boyd (6-10AM) switched shifts with Johnny Gilbert in November.

Scotty Day (10-Noon)

KCBQ RADIO AM 1170

Johnny Gilbert (Noon-3 PM) was known as Johnny G at KEWB in Oakland before he arrived to replace Jim O'Brien, who went to KBBQ in Los Angeles. He never caught on with the station management, preferring to go about his business in the tradition of his previous employers. He died in a helicopter crash in 1974.

Bobby Wayne (7-11 PM)

Larry Mitchell (11 PM-2AM) left to become PD at KYA San Francisco. Larry and Jack Hayes were close friends. Larry died of cancer in 1969 at age 33.

Jimmy Mack (11 PM-2AM) was a native San Diegan who graduated from Crawford High School. He briefly replaced Larry Mitchell in this shift in November before moving on to KKUA in Hawaii. This former lead singer in the rock group *The Measles* left an impression during his stint when he gave the staff mononucleosis. This was due to the staff's habit of sharing coffee cups and only lightly rinsing them when finished

Jack Vincent (2-6 AM) was beginning to close out the broadcast portion of his career. He was not only the station radio engineer, but a carpenter as well. He built most of the cabinetry and control consoles at the transmitter site in Santee.

Owner Lee Bartell spearheaded a move to relocate the station studio to the transmitter site in Santee. Planning for that relocation became the responsibility of GM Stan Torgerson. One day, while meeting with a telephone company representative to plan the phone installations at the new studio, Stan was presented a telephone bill for his review. He noticed that the phone in his office was generating calls far more costly than other lines in the station. The representative explained that the higher cost was due to an extension from his line to a receiver located at 2121 Upshur, the apartment of Phyllis Sandin.

By now, Bartell Media had installed Herb Mendelson as President of Bartell Radio Division. Stan Torgerson reported the matter to Mendelson. He, in turn, met with the Bartells and confirmed the extension to Phyllis' apartment with the telephone company. Since Phyllis was at the studio as calls were being forwarded to her home telephone, the only benefit from the extension would be having the calls taped.

David Bartell called Stan Torgerson that night and ordered him to fire Phyllis. When Stan noted that "Lee would be furious", David insisted that she be fired and that 'the board of directors will protect you.' Phyllis was fired the next day. A stormy meeting of the board of directors ensued in New York. David Bartell insisted Phyllis had to go. Lee countered that if Phyllis goes, so does Torgerson. The board caved in and agreed to the deal. Herb Mendelson was dispatched to San Diego to fire Stan. When Stan protested over the lack of having promised protection, Herb replied, "blood is thicker than water".

1968 – TEXAS MAFIA SNUFFS BOSS RADIO

KCBQ was on the move in several ways during 1968. An ownership and management shake-up occurred behind the scenes. Leonard (Len) Sable was hired as General Manager of KCBQ in January to replace Stan Torgerson. Torgerson moved into ownership of radio stations. Sable came from Meeker Radio Inc. where he handled Midwestern Operations. He fired Stu Langer and placed 'Big' Mike Scott as Program Director by March. As part of a merger deal with Downe Communications, Lee Bartell was replaced by Earl Tiffany as president of Bartell Media in May and he would never be a factor in KCBQ operations again. Len Sable was re-designated as Vice President General Manager in June so that the station would have a presence on the Bartell Board of Directors. Downe Communications would eventually assume ownership of the Bartell radio assets, although the long-standing Bartell name continued to often be used when describing the ownership.

The station's broadcast sound improved after moving to more spacious new studios at 9416 Mission Gorge Rd. in Santee. Broadcasts from the 7th and Ash studios produced an acoustic echo that was common in that era, especially as heard with broadcasts from powerful stations in Mexico. Better studio design and new equipment produced a richer sound at Santee.

There was an underground river at the tower site where sprinklers were used to keep the ground wet to enhance conductivity. Jack Vincent noted that a metal lattice mat was installed just below ground surface to accomplish the same purpose. Despite the groundwater condition, KCBQ, as the only 50,000 watt station in San Diego, was required to construct a studio-bunker 30 feet below the ground for civil defense purposes. The studio was never used, but did serve as a hideaway for 'recreational' activity. After continuous groundwater leakage, the studio was abandoned and sealed.

Mike Scott was now program director, music director, doing the afternoon drive, and putting in long hours as he heavily influenced the product on the air. The station rose to the top of local ratings in a highly competitive field and would remain there throughout the year while gaining national prominence. Mike applied his national industry contacts to locate and recruit new talent. Much of that talent had connections with Texas radio stations and became known as the 'Texas Mafia'.

KCBQ became known for program innovations. The station relaxed the typical format in the evenings to introduce listeners to alternative music with an edge. This began with DJ Bobby Wayne and continued with Jimmy Rabbitt. A sample of 'Funderground' music included non-chart 'progressive rock' music sections, such as Phil Ochs' *'Small Circle of Friends'*, Country Joe and The Fish's *'Feel Like I'm Fixin' to Die'*, Arlo Guthrie's *'Alice's Restaurant'*, Canned Heat's *'Amphetamine Annie'*, Velvet Underground's *'Heroin'*, Chamber's Brother's *'Time'*, Jimi Hendrix' *'Purple Haze'*, plus other Top 40 songs that fit the genre. This was Top 40 radio at its very best, although the music was hardly among Top 40's most popular songs.

'Melody Galore' was a young lady hired as a receptionist, (and whom Mike Scott swears is her real name), who became a public attraction, (see Bobby Wayne bio and photo), did personal appearances for KCBQ, and wrote a music column in GO Magazine. The 'Action' theme, with Action 30 survey and Action Men, was replaced on March 1 by 'Fun Lovin KCBQ with a Fun 30 survey in Go Magazine. The use of GO Magazine was optimized as a communication tool featuring columns by DJ staff and Miss Galore, recognition of news tips by listeners, the weekly survey, and contest information. Despite this, the station operated on shoestring budgets. Perhaps to make a point, the first contest involving 24 straight days of prizes, offered a shoestring as the first day prize.

The new staff lineup developed as follows:

Dex Allen (Claude Turner) (5:30-9AM) began his career as a high school DJ and got his first on-air experience at KLAC in Los Angeles as Chip Allen. He moved to KQV Pittsburgh where PD John Rook wanted to rename him Dexter Kilbride because he felt it sounded rich. Allen balked at the name, so they split the difference as Dex Allen. He was hired by Len Sable in early 1968 after Jim Carson turned down an offer to come over from KGB. He replaced Barry Boyd who took Johnny Gilbert's shift. Dex had been at KOL in Seattle. He prepared a weekly column on station activities in GO Magazine called Dex-tionary. He was temperamental early in his career, but mellowed over time. His shift suffered poor ratings against KGB. He went to KDAY in Los Angeles in late 1968 as Mike Scott assumed this shift. Dex returned in 1970 to work in the sales department through the mid-1970s. He now owns Commonwealth Communications and lives in the Rancho Bernardo area of San Diego County.

Scotty Day (9-Noon)

Barry Boyd (Noon-3)

Mike Scott (3-7 PM)

Jimmy Rabbit (Dale Payne) (7-Midnight) arrived from KLIF in Dallas in July 1968, where he had worked with Mike Scott. KLIF station PD Don McClendon had tabbed him as 'The (Jimmy)

Rabbit' to create the allusion of a Playboy image, and turned cars over at key Dallas intersections upon his arrival that read underneath "I flipped for Jimmy Rabbitt". Rabbitt was part of a group called *Positively 13 O' Clock* that had signed a record deal with Hanna Barbera. Coming to San Diego allowed him the chance to prove that 'album/underground' radio would work against KGB's *Boss Radio,* and allow him to be closer to the LA studios. He replaced Bobby Wayne when he was fired and went to WUBE in Cincinnati. Rabbitt described himself as being a hippie, anti-war protestor, and radio radical with trademark shoulder length hair and sunglasses. This was evident in his weekly column in GO Magazine, called RAP, ramblings about his personal experiences and album reviews. He served as the station music director and screened new album releases for the best tracks. Rabbit was highly respected for his ability to identify rising stars and had a great ear for music, with a preference for heavier 'underground' type records. 'Underground' was defined by Rabbitt as "new, fresh, and anything not being played at KGB." His ratings rose rapidly and he was lured away by KRLA in Los Angeles in February 1969 to apply his talent to a larger audience. Today, he is a song writer and 'musicographer'.

Lee 'Babi' Simms

(Gilmore Lamar Sims) (3-7PM) arrived from WGCL in Cleveland in July 1968 and took San Diego by storm. This applied to his colleagues as well who, while being interviewed for this story, wanted to discuss their experiences of working with Lee as much as talking about their own careers. He offered entertainment for listeners on the air and drama for management off the air. He held the philosophy that his employer needed him more then he needed them. To emphasize the point, he lived only in hotels with a few belongings, and always had a large new car always ready to leave town on a moment's notice. As a result, he is still considered one of the most traveled DJs in the industry. Lee loved to take people to the edge to see if they would walk off the cliff. It might be in the form of teasing, put-ons, or even testing, such as asking for money whether he needed it or not. Yet, one of his best pals was Scotty Day despite the completely different lifestyles each led. He never changed. He is quoted as saying that [1]"I live to be on the fucking radio. It is my only joy. When I'm allowed to paint my picture, using their canvas, their time, I bow down and say, "Thank you, gentlemen." Radio was his ego. He was very shy in front of an audience and low key off the air. On the air, he was considered among the most talented personalities to ever work at KCBQ. He had a unique gift of leading the listener through a verbal travelogue for an hour or more, all designed and perfectly ending with the introduction of a song. The following story does not have a song, but does offer a travelogue of Mr. Simms' antics spread over each year.

KCBQ and KGB had the means to monitor DJ banter in each other's studios both on and off the air. One day, the stations were competing to be the first to play a new Beatles song, but it could not be played before 4:00 PM due to record licensing restrictions, (and the delivery of the record). Lee's shift had not yet begun. As he awaited the record's arrival he considered how to build up audience anticipation to stayed tuned. The name of the song had not been released. So, Lee started thinking up song names and finally decided that 'Broken Down Old Merry Go-Round' sounded like a Beatles song for promotional purposes. Before he could get on the air, Bobby Ocean of KGB announced that he would soon be playing the new Beatles song 'Broken Down Old Merry Go-Round.' It was a rare day for Bobby to be had. But this was the day.

In 1968, Lee "Baby" Simms was 'drafted' from

the KCBQ studio. He had received a call on the air from a "Marine Corps" enlisted man that his "draft number" had been drawn and that he was being drafted. An ensuing ruckus occurred on the air as Lee was dragged away. Lee was screaming, "You can't do this to me", etc. as if he were exempt from being dragged, or drafted, into the service. This was on a Friday afternoon about 4:15 PM and he was bodily removed from the control room. Dex Allen took his place on the air and phone calls began to flood the station. This continued through the weekend until the following Monday afternoon. At 3:00 PM, when Lee was supposed to be on the air, Dex Allen was there until Lee showed up to reclaim his throne. He had set the whole thing up, and things quickly got back to normal.

Barry Boyd (Noon-3) remained on the air through Summer in this shift as the ratings declined. After leaving the air, he worked in the station sales department.

Gary Allyn (Noon-3PM) was on his way to a new job at KDAY after two years at KONO San Antonio. He stopped by the station to visit with Mike Scott who had worked weekends at KONO. Mike offered him the Noon-3 shift with the same pay that KDAY offered, so he began on Labor Day as a replacement for Barry Boyd. He became Program Director when Mike Scott was fired. Allyn was responsible for the station staff coming from KONO over the next year. Gary was 'Leader of the Little People" reflecting his 5'7"physical stature. He ran a Pat Paulsen for President campaign. As PD, Allyn kept raising the amounts of cash jackpots that caused the station to gain ground on rival KGB. This would prompt KGB owner Willett Brown to start the double cash jackpot campaign doubling the amounts KCBQ was offering.

Jack Vincent (Midnight-5:30 AM) had been ignored by station management for years as he did his broadcast from the Santee transmitter rather than the 7th and Ash studio. He did not follow formats mandated by a host of program directors. Mike Scott felt that the programming needed to be consistent 24 hours a day and removed him from the air in summer 1968. As a member of AFTRA and having a Class One operators license to work overnights, Jack continued to work as the station engineer (for another 15 years) because he was the only person at the station qualified to take transmitter readings and sign on and off the air log. A third class licensee, such as his successor on the air, B. Bailey Brown, could run the board and spin the records.

B. Bailey Brown (Bernard Bailey Brown Jr.) (Midnight-5:30AM) as a boy in the 1950's, Brown listened and studied the DJs on the radio and soon developed a love for broadcasting. He began his career at age 14 at KBER in San Antonio, TX in 1961. He worked for no pay in order to learn the trade. At 15, he was given his own show. After high school, he landed at KONO in 1962. Brown arrived from KONO during Summer 1968 to replace Jolly Jack Vincent in the overnight shift. 'B', as he was known, attended San Diego State College during the day, worked all night, and managed a few hours of sleep. He was part of the sweep by Bartell PD George Wilson to remove the Texas 'Mafia' and install a Milwaukee 'Mafia' in early 1969. 'B' worked for a few weeks at KPRI before returning to KONO. He moved on to advertising in 1988, then returned to broadcasting at KONO in 2001, where he also served as the station webmaster. This accomplished steel guitar player, who performed for years at night spots in San Antonio with a country band called Country Clover, died suddenly on October 24, 2003.

'Gentleman' Jim Carter (Weekends) began his radio career as a teenage driver of the KLIF Headline Cruiser in Dallas, TX. He joined the Marines and was hired at WHSL Wilmington, NC to do weekends in 1966. Jim was stationed at Camp Pendleton, after a tour of duty in Viet Nam, when he applied for a weekend gig at KCBQ and was hired by Mike Scott in March. Mike felt that Jim sounded like a southern gentleman on the air due

KCBQ RADIO AM 1170

to his accent. Jim found himself christened 'Gentleman' Jim Carter one day when he heard Scott introducing him on the air before his shift. The name has stuck over the years. After his enlistment was up, he returned to Dallas and was hired at KNUS. His career developed over the next 16 years, primarily as stations in Texas. He has been a telemarketer since 1986, currently working for San Antonio Wheelchair Athletes. He also tends to his hobbies of baseball card collecting, 1950s TV videos, and collecting comic books. His brief stay at KCBQ ranks among his highlights of his radio career.

Jim Hill, former Charger football player and current television broadcaster, did a 10:00 PM Saturday-6AM Sunday shift starting in November after torn knee cartilage preventing him from playing football. He emceed a show called 'Keep the Faith'. His radio career began in his native Texas where he did a radio program while attending college.

1969-1970 – ALLYN TYME

At the dawn of 1969, George Wilson replaced Len Sable with Dick Casper as Vice President General Manager. By now, the station had developed from no formula to a high energy and progressive format with a staff truly having a good time. With the programming freedom to play the best cuts from albums rather than just the hits, KCBQ had gained a ratings lead over KGB. George Wilson proved that arrogance and stupidity will overcome success. He directed PD Mike Scott to select a DJ on staff and fire them in order to establish authority. Mike selected himself and gave his two week notice. Wilson ordered Scott to leave immediately, and replaced him with Gary Allyn. Mike moved on to WKBK in Detroit, and was last on the air in 1988 and now owns a recording studio in Houston, TX that does national voice-overs.

More changes came to KCBQ in terms of programming and staffing as the Texans departed. GO Magazine reached its third anniversary in March and the station abandoned that publication and returned to the weekly survey produced by the station. The *Hit List* offered a more focused approach to promote the new staff. The format featured a 4 1/4 X 6 inch double sided folder that had a DJ photo on the cover. This survey format was no longer numbered, using only a weekly date. KCBQ held their ratings lead through the summer as substantial changes occurred among the air staff.

Jimmy Mack (Jim Talley) (6-10 AM) returned from KKUA Hawaii in January to take over Mike Scott's shift on an interim basis. He went to the 12-6AM overnight shift in June when Happy Hare returned. Jimmy's talents were in production. His air work was not bad, but Gary Allyn felt he needed to refine his skills in a smaller market and let him go. He started to leave without incident, but returned moments later and charged Allyn, grabbing him by the throat and choking him while others entered the room to pull him off. His exit from the station was the lasting memory he left.

Happy Hare (Harry Martin) (6-10 AM) returned to San Diego from Detroit to live in semi-retirement. He set out to write fiction and screenplays, while investing his earnings into real estate. He was soon recruited by Dick Casper to return to that station on the morning shift. As a publicity stunt, Hare agreed to jump out of a plane at 10,000 feet and parachute down into National City's Kimble Park at the start of a parade. After much practice, on the day of the event, he made his jump but opened the chute too soon. He drifted over Chula Vista as a convoy of parade watchers followed his flight. Avoiding a herd of bulls, he made a landing in someone's pool underneath the 75 lb chute. He was helped out of the water and made it back to participate in the parade in his soaking wet clothes.

Scotty Day had a shift change of 10AM-2PM

Gary Allyn (2-6 PM) wore many hats as the year began. First by having assumed the RAP column in GO Magazine (as defender of the 'little people') after Rabbitt left. Gary has always enjoyed writing, and he has written several plays. But writing the GO column was not among his favorite works. He became Program Director after Mike Scott left. He held the afternoon drive shift until Robert L. Collins arrived. He kept a shift at varied times on weekends through the year. He kept raising cash jackpots until the station resumed the lead in the market for a brief time.

The Milwaukee Mafia began when George Wilson continued the tradition of transferring DJs among the Bartell stations. He brought in two jocks from station WOKY in Milwaukee:

Robert L. Collins (2-6PM) was from WOKY Milwaukee, but arrived via KFI Los Angeles in July to take the late afternoon shift. He was with the station about a year and provided a ratings and talent boost with his Robert L. Collins Extravaganza show. He timed record intros for talk-overs with a ten second sweep stop watch. Digital timers would later be standard equipment in control rooms, but in this era, most DJs were doing talk-overs using instinct alone. His southern drawl and sense of humor made him a favorite among mature ladies. He then returned to Milwaukee at station WRIT. But he was best known at station WGN Chicago where he reigned over the airwaves for decades. Bob loved cars, motorcycles, cigars, and beer. He died in an air crash in 1999.

Ron Thompson arrived in March 1969 from WOKY in Milwaukee to replace Jimmy Rabbitt in the 10 PM-2AM shift. Ron developed the apt nickname of 'Ugly' that reflected his relations with other staff. He also expressed little knowledge about the community on the air. He went to the 6-10 PM shift in February after Neilson Ross arrived Ron was last heard from in the late 1990s when he called Bob Collins from a pay phone at O'Hare airport in Chicago explaining that he had married a Thai woman and was heading back to Thailand where she owned a hotel.

Mal Harrison (Noon-3) is a name only a former PD would be able to recall. Having been hired by George Wilson he arrived at the same time as Collins and Thompson. He worked for two weeks, and was gone without a word, never to be heard from again.

Neilson Ross (Neilson Ross) (10-2AM) arrived from KKUA Hawaii in February and initially worked on production and weekend shifts. Neil had gone to high school in San Diego and was anxious to return to a local radio station. He asked Jimmy Mack to bring an audition tape to PD Gary Allyn when he was hired in January. From that tape, Gary extended an offer to Neil. He took the 10 PM to 2 AM shift by April. He was born in England and raised in Canada and Southern California. He took the overnight shift after Jimmy Mack left in November.

Lenny Mitchell (Leonard Shantzer) (Midnight-6 AM) arrived in May from KEZY in Orange County to work weekends and do production work. He had been on the air under the name of Lenny McGuire at KLAN Hanford  where all the DJs, (including Mike McGregor of KDEO) used 'Mc' as part of their name to differentiate their clan from another Klan. He moved to KAFY Bakersfield in 1966 under PD, and *Radio and Records* Magazine founder, Bob Wilson. Lenny selected the name Lenny Mitchell. Lenny soon struck up a friendship with the DJ on the preceding shift, Lee Simms.

Lee 'Babi' Simms had gone to KONO, KTSA San Antonio, and WJBK in Detroit in 1969, (where he was reunited with Mike Scott). He returned in September to resume his show at 6-9 PM. Lee opened his show one night with a musical set instead of his usual chatter. Then he played ads followed by another set. Finally he announced that he was feeling down because he got word his mother died. He was trying to reach his sister to confirm but the phone lines were tied up. He would place calls on the air and commiserate with the operator on how depressed he was. Finally, he got through to his sister who confirmed that their Mother was lying in the next room. He said, "Sis, put the phone down by Mom's ear. Somehow I just know she'll be able to hear me." Suddenly she says, "Lee, Is that you?" He answered, "Thank God your alive! I thought you were dead!" She replied that she, "might as well be dead as often as you call" . . . on comes *Gotta Get a Message to You* by the Bee Gees.

He often got into trouble for disparaging remarks about sponsors that would end up losing revenue to the station. When Lee followed a Beneficial Finance ad with an admonishment to his listeners to 'stay away from these bandits', the finance firm called GM Dick Casper, who called Gary Allyn to fire him immediately, which he did then and four other times. Another DJ had to come in to finish that shift. Lee would later return, again and again.

The station produced an album of 22 Heavy Hits in November 1969. Kay Kaiser drawings of all the staff appear on the back. Kay had grown up in the Milwaukee area and was familiar with many of the staff now at KCBQ. She was the only child of aging parents who moved to San Diego as she attended college. Graceful, well-educated, and a talented artist, Kay drew portraits of the staff, typeset the play list, and handled the printing for the weekly surveys. The survey was renamed in October as the Silver Dollar Survey with a Peace Dollar logo to promote their Fall contest. All other format features remained the same.

Two other jocks were on staff for a brief period at this time. Shifts were realigned to add a position.

Thom Devine (Weekends) had worked with Gary Allyn at KONO San Antonio. 'Texas' Thom followed Gary out to San Diego seeking a job to earn money while getting settled into something else. After leaving KCBQ in 1970, he did overnights at KFXM San Bernardino in 1971.

Joe Light (Noon-3) was a George Wilson hire who arrived from KRIZ Phoenix for a brief time in the fall following a lengthy career in the Midwest. Shifts had be realigned to accommodate a new Noon-3 shift between Scotty Day and Robert Collins. Heavy drinking made him loud, brash, foul mouthed, and unreliable. He was divorced and had custody of his two daughters ages about eight and twelve. He left them with Paul Oscar Anderson to care for them when he left town. He did not return, and Paul eventually put the girls on a plane bound for their father. Joe was later involved in radio-related sales.

1970

By this year, the fortunes of the Fun Lovin crew were being overtaken by the Boss Jocks at KGB. KGB's Buzz Bennett was authorized to double the cash jackpots over the amounts KCBQ was offering. However, KCBQ's Dick Casper was restricting cash prizes in order to produce good annual bonuses. So, Gary Allyn developed innovative programming to keep an audience. A talk show program began on Sunday nights. The weekly survey was renamed in May as the *Fun 30 Survey* with a sun logo. The promotions shifted from contests to concerts. All other format features remained the same. The survey was renamed again on November 27 as the *Long Play Survey* as the emphasis was shifted to album play. A list of top albums appeared in this format.

Happy Hare (5:30-9 AM) remained on the air through 1970, but left when the 'hipper' culture that Bennett brought in clashed with his style. Hare was interested in radio management and knew that the management path was through sales, not air work. He joined the station sales department, then moved over to KSON with Sales Manager Mike Stafford. Hare was offered a job to host a morning news program at KSDO by Peter Lund, but turned it down. Lund later became CEO of the CBS network and launched Hare in a new career in TV voiceovers and writing. He returned to the air from time to time, such as in 1978 when he anchored the Walk for Mankind Day. After 30 years off the air, Clear Channel recruited Hare to work a weekend shift at KPOP, that was formerly the dial location of rival station KGB. Hare has since moved to the morning shift in a free form format reminiscent of early Top 40 radio.

Scotty Day (9-Noon) went to KOGO in May after 12 years with the station. He could have cost the Bartells their FCC license because he was never paid for the production work he did early in his career. However, Scotty loved to play the stock market using Bartell stocks he received in lieu of cash bonuses. His stock strategies made him a wealthy man, enjoying life and his Buick convertible by the time he left the station in 1970. We're left to speculate on why he left KCBQ at this time after so many years. The music was certainly moving away from his genre. He had several buddies at KOGO that included Jerry 'Shadoe' Jackson, Don Howard, and Ernie Meyers. The arrival of Dick Casper as GM may have signaled changes to come. He stayed at KOGO until that station was sold in 1979 and had succeeded Jerry (Shadoe) Jackson as program director. Scotty was a private person who shunned celebrity, but was dearly admired. He died on October 4, 2002. A prolific pipe smoker, the cause of death was from complications of pneumonia derived from lung cancer.

Gary Allyn (9-Noon) returned to the air in May when Scotty Day went to KOGO Radio. He left in December upon the arrival of Buzz Bennett, who soon called to thank Allyn for delivering a station that returned to the top of the ratings the previous fall. Privately, Buzz would confess that "Allyn left me a good Book, man." If one looks at the talent he had, (and had to manage), the aggressive contests to counter Boss radio, the program innovations into talk radio and late evening funderground, and the ratings he produced in spite of the Drake Radio phenomena of the late 1960s, there is ample evidence that his work was not adequately recognized and rewarded. Gary was at KDEO in 1973 and PD of Top 40 KSEA-FM in 1974. But his substantial career, which included air work, production director, program director, music director, operations manager, ownership of Top Spots recording studio, ownership of LYN records, and development of the radio comedy serial adventure OB Ranger, and Play-by-Play Sports Fantasies, no longer included running a radio station. Gary now lives in Fallbrook.

Chuck 'Magic' Christian (Charles Christensen) (9-Noon) arrived in August when Gary Allyn returned to Program Director duties full time. Chuck had worked under his real name at KMEN, then Chris Charles at KGBS. Chuck acquired the nickname 'Magic' when the movie  *The Magic Christian* was released in 1970. Magic had the talent to be great when he wanted to be. But at this time, he was inconsistent and would sometimes wing his way through a show rather than be well prepared. Sometimes his attention would be focused on talking to a girl on the phone while on the air.

Neilson Ross (Noon-3) moved to this shift when Lenny Mitchell took the regular overnight shift. When new PD Buzz Bennett arrived at the station, Neilson was fired at the end of his shift because Buzz felt he didn't sound positive enough on the air. He later joined Gary Allyn at KDEO. Neil left radio in 1985 and now does voice-overs and impersonations for products and media networks.

KCBQ RADIO AM 1170

K.O. Bayley (Bob Early) (3-6PM) arrived in May from WCBS-FM in New York when Robert L. Collins left for Chicago. This shift was soon adjusted to 3-6 PM. K.O. was formerly a KGB Boss Jock under the name of Bob Elliott in 1966-67. He had been a professional boxer under the name of Jackie Joy. His name was derived from K.O. for 'knock-out' and Bayley as tribute to his bay area roots. He never drove, possibly due to poor eyesight, so his girlfriend Pam shuttled him everywhere. He would return to KGB later in the year, then follow Gary Allyn to KDEO.

China Smith[2] (Tom Rohrbacher) (6-9PM) arrived from Seattle where he used the air name of Wayne Thomas, replacing Ron Thompson. Upon his arrival, Program Director Gary Allyn encouraged him to use the name 'China'. Thomas was already a student of Oriental philosophy and was intrigued by the idea, but felt it should be anglicized in some way. 'Smith' sounded all-American so the name 'China Smith' was born literally moments before he went on the air for the first time in San Diego. He had the goal of eventually landing in the Los Angeles market, particularly at station KRLA. Shadoe Stevens hired him from KDAY in 1971 to come to KRLA.

Lee 'Babi' Simms (9-Midnight) retained this shift as a means "to avoid corporate assholes" and provide him the freedom to smoke a joint or drink a beer while on the air. Lee brought some brownies his girlfriend Annie had made for his birthday and left them for the staff to enjoy. It turned out the brownies contained hashish. Lee was soon recruited by Dick Saint, a former KGB DJ, and current KRLA Program Director. Dick invited him to come up to take the same shift in a larger market.

Lenny Mitchell (Midnight-6 AM) was developing as a reliable steady overnight DJ. He was christened as the 'Midnight Rambler' by Lee Simms.

Phil Flowers (Phil Boles) grew up in Oceanside, CA as a listener of KCBQ, and would ride his bike to local station KUDE to learn all he could about radio. As a youth, he would urge family or friends to drive him to downtown San Diego. There, in the picture window of KCBQ studio at 7th and Ash, he would watch radio

personalities perform on the air. Other times, he would visit with Jack Vincent at the transmitter in Santee. Phil enrolled in broadcasting school as soon as he graduated from high school. He was drafted in the Army in 1965 and served in the same unit with TV news anchor Harold Green, who subsequently became program director at KDIG, and hired Phil. Phil later moved on to KOWN in Escondido, that had a modern country format. He convinced management to do a Top 40 format at nights and worked himself into a nightly show while serving as the night PD. During this time, he worked with Dex Allen, of the KCBQ sales department, to book groups at Palomar College for concerts. Dex encouraged Phil to come over to KCBQ, and Phil started doing weekends at KCBQ as a replacement for Thom Devine. He continued to work weekday nights at KOWN as Phil Boles while working weekends at KCBQ as Phil Flowers. Gary Allyn helped him develop the name Flowers to emulate 'flower power' which was in vogue at the time.

He grew up as a KCBQ listener, befriended the staff of the station over the years, became a DJ at the station, and was there at a pivotal time in station history. It's no wonder that he held a connection with the station long after he left, and throughout his radio career. Phil compiled an extensive library of audio archives, developed a wealth of history about the station, and shared that information on a weekly program on KCBQ called "I Still Q In My Car" that ran from 2000-2003. Phil Flowers became the resident historian of KCBQ radio. He died of a heart attack at his home in Oceanside on December 22, 2003 at the age of 58.

Paul Oscar Anderson (Paul Brown) (Sunday 9PM-12:30AM) became the station news director after retiring as an announcer with NBC, and touring the country with his wife. He had exhausted his funds and

needed to resume an income. He had a voice best described as "God-like" that was used to host of a talk show called *Talk Right In*. The title was a take-off of the song 'Walk Right In'. The show was a mix of topical discussion and off-beat psychic theory that involved a co-host named Dromedus. Callers would share their perceived psychic experiences and Dromedus would interpret them. Former Charger player and current sports newsman Jim Hill also contributed to the show following his own *Keep the Faith* show. Paul was often referred to as POA off the air. He was also on the air in Denver and Chicago markets. Like others in radio, he drank too much, but unlike many, he got control of it and remains sober.

The station moved into an emphasis on album formats by the end of the year. The perception was publicly held that the station was in a downward spiral in ratings that only turned after Buzz Bennett arrived. The fact is that the album format caught on in the few months it was in place and KCBQ was on top during Fall 1970.

THE NEW 'Q'

After struggling with declining resources, Vice President General Manager Dick Casper scored a coup in December 1970 when he signed Buzz Bennett to program the station. He had already been named national Program Director of the Year in 1968 for turning WTIX in New Orleans around in their market. Buzz Bennett put KGB back on top of the ratings in San Diego. He became available to the station when RKO, owners of KHJ, refused to place him as program director at KHJ based on his non-conforming appearance and lifestyle. Programming KCBQ would be a keen act of revenge. He radically changed the staff by firing KO Bayley, Neilson Ross, and China Smith. Happy Hare was also terminated, but quickly moved into the sales department (and better income). Lee Simms was on his way to KRLA. Buzz retained Chuck Christian, Lenny Mitchell, and Phil Flowers. Always popular with the air staff, he then brought his former KGB jocks over that included Bobby Ocean, Rich Brother Robbin, and Harry Scarborough. They were followed within months by Christopher Cane and Shotgun Tom Kelly. This gave rise to his first major contest when he had his staff remain nameless and have listeners guess who they were. The winner would get a new Mercedes Benz used in the Janis Joplin hit to frame his programming. He programmed around other songs, such as the Seven Days of Maggie May, where the listener had to count the times the song was played over a seven day period to win big bucks. *The Great Ripoff* was his next contest, perhaps to celebrate ripping off the KGB staff.

Buzz introduced many successful innovations destined to be copied by others. He introduced shorter jingles that dropped the reference to 1170 kilocycles, but added the name of each disc jockey into a jingle. This included the 'shotgun jingle' with just drums and call letters, nothing else. This has become an industry standard. He speeded up song tracks to make the competition sound slow and draggy. He also kept songs longer on the chart that demonstrated extended listener appeal, rather than rotate them off automatically. He inaugurated uninterrupted music sets, in the tradition of FM radio and demanded high energy enthusiasm by DJs to notch up the excitement for the listener. An entirely new survey format was created with Bobby Ocean art work and a numbering series starting at Issue I to emphasize a new beginning for the station.

One of Buzz's most significant contributions was the now-generic concept of DJs announcing the call letters **first thing** every time they spoke. This came about almost by accident right after Buzz and company arrived in January to fire up the 'New Q'.

One of the linchpins of Buzz's programming philosophy was forward momentum, "curtain always going up." To that end, he was constantly seeking ways to streamline the flow of sound elements from one record to the next. One day, about a year earlier, when Buzz, Rich Brother Robbin, and Bobby Ocean were still at KGB, they were together in a car and pulled up at an intersection where a young guy had the radio turned up loud listening to KGB in the next car. The song ends and Gene West comes on in the backsell, "The Rolling Stones and *Honky Tonk Women*, good afternoon this is Gene West . . ." But by this time, the kid has turned the volume back down without hearing the call letters.

Now its 1971 at the beginning of the 'New Q'. Buzz went to the studio and says to Rich Brother Robbin, "Hey, remember the kid at Midway and Rosecrans who turned down the radio before we announced the call letters? Wouldn't it make more sense to bark 'em out and burn 'em into the listener's brain *before he could tune out?*" Rich tried it out and Buzz was so happy with the way it sounded that the station began using the concept immediately on an 'every-time-you-open-your-mouth' basis. At the same time, Buzz had Rich experiment with starting the next song in a sweep before talking, (instead o talking over the tail of one song and the intro of the next). These two ideas, call letters first, plus starting the next song before talking, have become radio format standards lasting to this day.

Buzz introduced shock music that was popular with a teenage audience. Examples included *DOA* by Bloodrock about a fatal accident, and *Timothy* by the Buoys about a dog, that seemed like it was about cannibalism.

Buzz also encouraged his staff to always answer the request line, even though, being programmed, they were not going to play the requested song. But he felt that since the listener took the time to dial up the station, the staff should talk to them because each caller represented a rating point, and ratings were very important to Buzz.

During his one year tenure, The King of Contests set a program format that would remain for the decade while other AM Top 40 stations dropped out.

Buzz also had an innovative plan to inflate ratings. The station printed booklets that looked like Arbitron Ratings diaries and informed listeners that if they received one, to fill in KCBQ as their listening choice and return it to have a chance to win $1000 in a drawing. The station expected to collect both real Arbitron booklets and booklets printed by KCBQ from listeners, and that the contest would increase the number of true Arbitron diaries returned with KCBQ filled in. This contest had two results. It eliminated rival KGB, Bennett's former employer, from the Top 40 market. But it also led to changes in Arbitron rules that remain in effect today.

The 1971 'Q' Crew:

Chuck Browning (6-9 AM) a native of Memphis arrived from KRUX in Phoenix. 'Chucker', as he was called, opened the 'New Q" format but would be gone by July when Buzz Bennett fired him for playing a cut from a Beatles album he  specifically told him not to play. Chuck would resurface at KGB in September, then return to KCBQ to take the 12-3 shift before the end of the year when Chuck Christian left. Chucker was famous for being late to events, or not attending at all. This applied to even those he was suppose to host. He would often offer an incredible explanation summing up with 'That's my story and I'm going to stick to it.' Browning was a heavy smoker and toker who died of lung cancer on March 3, 1988.

 Peter Huntington May (9-Noon) arrived from KGB and held a close association with Buzz Bennett, that extended to joining him at other stations he programmed.

Chuck 'Magic' Christian (Noon –3), a survivor from the 1970 crew for a few more months. He went to CFTR in Canada in fall 1971. He ended up doing three stints with the station before being fired in December 1974. At that time he went to CHUM radio in Toronto. He was with KLCY Salt Lake City in 1990.

Bobby Ocean (Ernest Raymond Lenhart) (3-6 PM), the signature voice of KGB now doing the same for KCBQ. But after 10 years in radio he was getting disillusioned over the whole concept and quit on June 18 after airing a farewell address to his listeners. He returned to KGB by October as Program Director and did the AM drive shift.

Rich Brother Robbin (Richard Werges) (6-9 PM), a former KGB Boss Jock returned from Detroit and functioned as Assistant PD, music director, as well as the evening DJ. He writes that, "the single most amazing thing I've participated in during my 44 year career was the major event in the evolution of KCBQ's lightening fast formatic movement." Rich served as station program director in the 1970s and again in 1996. He is now at WGFX in Nashville.

Harry Scarborough (9-Midnight) arrived from KGB and brought his infamous free spirit approach to life. Harry moved to the 6-9 AM shift in July

Lenny Mitchell (Midnight-6 AM) survived as a result of Buzz Bennett's appreciation of his consistent air work and because he held the overnight shift.

In April 1971, Bartell bought WMYQ in Miami, FL and several staff helped set up the station format either at the site or with production work based in San Diego. Bobby Ocean assisted in the production work.

 Phil Flowers (Weekends) moved to WMYQ to help Buzz Bennett set up the station. After Buzz left, new management wanted Phil to leave so they could bring in their own people. Phil rejoined Buzz and Chuck Browning at KUPD in Pheonix in 1972. In 1975 he was hired by Bobby Rich as a charter DJ with B100 FM in San Diego. He returned to the 'Q' from 1978 to 1981 to do production work, traffic reporting, and as the weekend DJ. He returned again in the 1990s.

By July, several more changes occurred among the staff:

Peter Huntington May (9-Noon) moved on to WMYQ. He ultimately went into station ownership in the Duluth, MN area that failed. Later he did sales work for a company owned by Rich Brother Robbin in Tucson, AZ. He does not appear to still be in radio.

Christopher Cane (Kenneth R. Witt) (9-Noon) arrived from KGB August as fellow former KGB Boss Jock Peter Hunting May followed Buzz Bennett to WMYQ in Miami.

Dave Conley (9-Midnight) arrived in July when Harry Scarborough replaced Chuck Browning in the morning drive shift after he was fired. Dave had worked with Shotgun Tom Kelly at KACY Oxnard, where he was nicknamed the "Clean Living Kid', and KAFY in Bakersfield, and was directly involved in developing the name of Shotgun Tom Kelly.

Batt Johnson (Weekends) arrived from KUPD Phoenix to replace Phil Flowers beginning in July, and had the distinction of being the first black DJ and a valued 'entrepreneur' at the station. He moved on to KPRI-FM. He now runs a communications consulting firm advising personalities and companies how to promote themselves in media. He has appeared as a VJ on VH-1 and in movies and television.

Shotgun Tom Kelly (Tom Irwin) (Noon-3 PM) had quit his job at KGB rather than accept a trans-

KCBQ RADIO AM 1170

fer to KYNO Fresno. He got a call from Rich Brother Robbin, inviting him to come to KCBQ. Peter May and Bobby Ocean had left. Chuck 'Magic' Christian moved to the 3-6 PM shift to replace Bobby Ocean, so they needed a Noon-3 PM DJ. Shotgun had come to San Diego from KAFY in Bakersfield in February 1971. His tenure at KGB was his big break into the business. This would not be his only tour with KCBQ or KGB.

A remarkable story behind Shotgun was that he grew up in San Diego and had a paper route in the area of the old 7th and Ash studio. The public was forbidden from entering the studio since DJs could be watched from the street through a second floor glass window enhanced by a mirrored ceiling. Tom HAD to see the studio from the inside. One day he went in and asked for Scotty Day. Scotty came out and Tom explained he was doing a class project on vocations and asked if he could photograph the studio. Scotty let him in to take a rare surviving photo. He would visit the KCBQ transmitter late at night where Jack Vincent would let then in to watch the show. He visited a remote KOGO studio at Oscars restaurant in Lemon Grove one day and Frank Thompson interviewed him on the air. That's when he knew he wanted to be a DJ. Tom met his future wife Linda at KPRI where he worked weekends. They were reunited at KCBQ where she scheduled ads.

The station cut a deal with Apple Records for exclusive rights to air a new John Lennon single for the first time in the nation. An acetate of the song arrived by jet and reached the station during Shotgun Tom's show. Shotgun quickly set up the piece and announced 'Here's the new single from John Lennon . . . 'Imogene'. . . as the words 'Imagine there's no Heaven . . .' began to play. He corrected the title as the song ended, and most listeners probably never noticed.

After ten months from inaugurating the new format, only Rich Brother Robbin and Lenny Mitchell kept their shifts.

1972

And then there were none. Since adopting the top 40 music format in 1955, KCBQ was always in hot competition with other top 40 stations. First, the Mighty 690 XEAK created competition in April 1957, then dropped the format in May 1961 when Gordon McClendon bought the station and introduced an all-news format. KDEO broke in during 1959 but dropped the Top 40 format in March 1966 to do 'all requests', followed by oldies, then album rock in the 1970s. KGB began their effort in 1963, perfected it by 1965, and swapped ratings leadership with KCBQ until April 1972. Now KCBQ had the AM Top 40 market to themselves. Despite this tremendous advantage, there was means to screw it up.

GM Dick Casper was in charge of setting up WMYQ. Something went wrong and he quit. On January 4, 1972, Buzz Bennett followed him out in protest and both went to work for Heftel Broadcasting. Several of the ex-KGB DJs, including Rich Brother Robbin, Harry Scarborough, and Shotgun Tom Kelly left, and so did the news readers. Chris Bailey and Gary Price were flown in from Bartell station WOKY in Milwaukee to work alternating four-hour shifts until new staff could be hired. Hap Trout was brought in as General Manager to replace Casper.

Jack McCoy replaced Buzz Bennett as Program Director. Jack had been program director at WAPE Jacksonville, FL. McCoy arrived to find no staff and a depleted budget resulting from aggressive contests. KGB had moved up in the ratings so a new foundation had to be laid. McCoy set out to hire staff who 'had a vocabulary that would appeal to the youth on the beach.' Jack and Doug Herman then set out to beef up the station contest package without bankrupting the station. Mardi Nehrbass served as Music Director. They teamed with staff Carol Craig, Sherri McCoy, and Chris Cane to devise the legendary *The Last Contest*. KGB and KCBQ had continued to outbid each other with prizes until the Last Contest was devised where listeners could call in and win prizes by package number rather than a stated prize. Therefore, the

KCBQ RADIO AM 1170

competition couldn't up the ante. Prizes included everything from a trip to Sea World to a new home.

The contest reached a climax with everyone phoning it at the same moment. Unforeseen was the effect of blowing out the phone system in the metropolis of San Diego. The contest was enormously successful and went into syndication. Ratings soared as a result. This would lead to better things for Jack McCoy and Doug Herman. Hap Trout got the station into some FCC trouble over advertising spot sales and was replaced by Russ Wittberger.

Charlie Tuna (Art Ferguson) (6-9AM) is more closely associated with the LA station market, most particularly KHJ, where he had replaced the legendary Robert W. Morgan in the morning drive shift. When Morgan returned in 1972,

Tuna quit rather than accept another shift. His career in radio reaches back to age 16. The name Charlie Tuna arose from a Starkist tuna ad while he was at KOMA Oklahoma City in 1962. Tuna arose at 3 AM Monday through Saturday to commute from his Tarzana home to Santee to do his 6 AM show each day. He had to remain in the LA market to work on syndicated radio programs and do a shift on the Armed Forces Radio network. But he had to work outside of the LA radio market at this time because of a non-compete clause in his KHJ contract. Station manager Jack McCoy asked him to come aboard during the Spring rating period knowing he had committed to a new deal with station KROQ that would begin in the summer after the KHJ contractual problem resolved itself. He could create a sensation wherever he went by having celebrities call in, or calling Mattell Corp. for a date with Barbie. His show was a market ratings leader during his brief tenure. He continues to broadcast on KBIG.

Christopher Cane (9-Noon) moved on to KYA in San Francisco where he worked a shift and served as PD. In 1979, he returned to KGB when it was 13K. He is no longer in radio and resides in San Diego.

Danny Martinez (Noon-3) arrived at the beginning of the year and stayed until June when he was hired by the Drake organization at WOR-FM in New York City. He stayed there for a year before going to KHJ Los Angeles. Danny can now be heard on Arrow 93 FM in Los Angeles. Lenny Mitchell followed him in this shift briefly

Don Fox (James Warren Rowley III) (3-6 PM) was president of the fan club for DJ Skip Bell in Grand Rapids, MI while performing in the Jay Hawkers band. He got his first air experience playing tapes at WKNX, located in a Victorian home in Saginaw, MI. He moved on to Detroit as Jerry Baxter and to CHUM in Toronto as Johnny Williams. He heard about KCBQ in San Diego when Chuck 'Magic' Christian joined the staff in January 1972. He made a cold call to Jack McCoy that led to his hiring. He took up residence literally around the corner from Buzz Bennett, who was best man at his wedding. McCoy wanted a quick two-syllable name to use on the air so the name Don Fox was born. He took the afternoon shift that Magic Christian held just months before.

Dave Conley (6-9 PM) used a surfing vernacular that held particular appeal to the beach crowd that McCoy was seeking.

Chris Bailey (9-Midnight) had arrived from WOKY in Milwaukee in December 1971. Chris was anxious to return to his family in Milwaukee, so he tried to be obnoxiously loud on the air to wear out his welcome. Unfortunately for his cause, this was just the sound McCoy was seeking. Bailey returned to Milwaukee briefly, but came back on the day Charlie Tuna arrived. He took the night shift to become a very up-tempo personality. He had a particular knack at selling a song to the listener, particularly Rolling Stones music. He would intro a song as if by surprise that it was his favorite, and he hadn't heard it in years, and convey that excitement to the listener.

Lenny Mitchell (Midnight-6AM)

KCBQ RADIO AM 1170

Dave (the Snake) London (Dave Cockrell) (Weekends) replaced Bat Johnson. Mike Wingfield had been asked to take this assignment but backed out due to family problems. McCoy had worked with Snake at WAPE in 1970 and brought him in. Snake remained on staff for about eight months. He is now an electrical engineer in Knoxville, TN.

Through a variety of circumstances, the station was short on DJs by summer. Charlie Tuna returned to the LA market as soon as the KHJ non-compete clause had lapsed. Danny Martinez, moved on to WOR-FM. Bill Moffitt, Matt Guinn, and Gene Knight were all hired in the same week. The new lineup as of August was as follows:

Don Fox (6-9 AM) left with Charlie Tuna for KROQ in Los Angeles, but when he realized the station had a very weak transmitter, he asked McCoy if he could return. Fox did and took the morning drive shift. Don could not pass up the chance to work the LA market at KHJ in November 1972. Upon arrival, Robert W. Morgan felt the Don Fox name was pedestrian and suggested the more Californian name of 'Sonny Fox', by whom he is known today.

Bill Moffitt (9-Noon) arrived from KJAE Denver in the summer after sending an aircheck to Jack McCoy.

Matt Guinn (Noon-3 PM) arrived from WIXY Cleveland in the summer and was on staff a few months. Though he had a big announcer-style voice, he was thought to have some short-comings in production, originality, and content. He did have luck as he and his wife appeared on the Newlywed Game show and won furniture.

Bobby Noonan (Drew Harold) (Noon-3 PM) arrived in September from KISN in Portland, OR. He quickly struck up a friendship with Gene Knight, then just as quickly left for Boise, ID where he remains today on station KBOI. He did the 9-Midnight shift after Chris Bailey left.

Dave Conley (3-6 PM) continued on, but at a new time.

Gene Knight (Jerome Peterson) (6-9 PM) knew in junior high school that he wanted to be a DJ. Living in Connecticut at the time, he felt WMCA, the leading Top 40 station in New York, served as a model format with great personalities. He went to high school in Escondido where KOWN offered a middle of the road format during the day and Top 40 at night. He set his sites on landing a job there, and did upon graduation from high school. But after three months, the station went country full time. He was using his real name of Jerry Peterson and followed PD Mike Larsen to KSON, the country giant in San Diego. In 1971, he had the opportunity to join the staff of KSEA, and PD Jerry Clifton wanted his new guy to adopt a new name to shed the country DJ identity. Gene Knight was the name Clifton selected from a list Jerry had prepared. Gene got a call from Jack McCoy in summer 1972 with a job offer to come to KCBQ. Gene wanted to resume his real name, but Jack felt he was too well known as Gene Knight, so the name stuck. He arrived from KSEA in August with his Gene in the Knightime show.

Chris Bailey (9-Midnight) describes KCBQ as his best radio experience. "There was always excitement at the station. People knew they were the best. It was a magical time and a magical station." Bailey was lured by George Wilson to set up station KBEQ in Kansas City where he served as PD. That station produced a KCBQ-type sound. Bailey eventually moved on to perform voice tracks, but wanted to use his masters degree in psychology professionally. He has done so as a private investigator.

Lenny Mitchell (Midnight-6AM) had replaced Danny Martinez in the Noon-3 shift, but returned to overnights when Matt Guinn arrived. Lenny was wrapping up his long tenure with the station. He would soon be moving on to KSEA. He maintained that "The DJ's job is to make the audience feel good about music. The artist is the star. The DJ's enhance the music. The DJ and the station that does that consistently become No. 1." Lenny has also been an avid collector of baseball cards.

1973

Jack McCoy had begun syndicating *The Last Contest* package to stations all over the country and received funding from Downe Communications for research and development into computer technological applications. He teamed with Doug Herman to develop programming and contest packages as their new company occupied space at the studio building until 1975. They returned to the building in the early 1990s, and were the last occupants in 2003.

Rich Brother Robbin took over as Program Director on July 1, 1973 for his second run at the station and kicked up the high energy sound of DJs and promotions. Rich had been working with Don Barrett as PD and format architect of KFOX-FM in Los Angeles.  KFOX was the first FM station to promote itself with the dial location (K-100), as it changed its calls to KIQQ. Rich placed an echo button on the control board that had been removed in 1971 as the New Q emerged, to the delight of the DJs. Rich has a good ear for music and recognized the residual benefit of the *American Graffiti* fad at the time of charting certain oldies with current hits He chose not to delegate music selection to a Music Director. Rich writes that, "NO PD IN HIS RIGHT MIND would farm out the music to a subordinate. It is way too important. I've carried that philosophy to this very day. I have never let another person do my music without my direct supervision and approval of the day's programming." Rich also tended to hire high profile DJs who could dispense energy and enthu-

siasm over the air. Russ Wittberger remained as General Manager through the 1970s.

The 1973 staff included:

Bill Gardner (6-9 AM) replaced Don Fox in the AM drive when Fox went to KHJ in Los Angeles. Bill now does mornings at KOOL in Phoenix.

Chuck 'Magic' Christian (6-9 AM) returned for a second run with the station. He was considered a great set up man to do anecdotal stories on the air causing listeners to call in and comment. One example was a story about his lunch starting to smell funny and whether or not he should eat it.

Bill Moffitt (9-Noon) would not only be a fixture at the station until 1976, but would return in 1981 when the station made the transition to KCBQ Country.

"Shotgun" Tom Kelly (Noon -3 PM) returned for his second tour with the station in the Spring. He had followed Buzz Bennett to KRIZ Phoenix, but got homesick for San Diego. Dave Conley notified him that Matt Guinn had left so Tom contacted Jack McCoy who brought him back in.

Dave Conley (3-6 PM) Conley was dating Sue Robinson, the secretary for Russ Wittberger in 1973 and she posed with the DJ staff on a year-end survey cover. Dave had a legendary reputation for heavy use of alcohol that led to the end of that relationship. It was rumored that he died in the summer of 2001 in Texas, but there are further reports that he is alive and well.

Rich Brother Robbin (6-9 PM) took over this shift.

Gene Knight (9-Midnight) replaced Chris Baily in this shift.

Dick Young (Midnight-6 AM) had known Jack McCoy elsewhere and contacted him about a job while attending law school. Jack opened up a position by firing Lenny Mitchell on January 19 to create the slot for Young. Dick held a Class I license to handle the overnight engineering duties and to provide a smooth transition from the loud party-style shifts that preceded. As for Lenny, he joined Gary Allyn at KSEA, then moved on to KFMB AM. He left radio in 1977 to begin a 24-year career with San Diego Transit Corporation. Lenny now lives in La Costa.

Beau Weaver (Beauregard Rodriguez Weaver) (Weekends) was working at KNUS in Dallas, TX under PD Michael Spears when he received a job offer to come to WEFM Chicago. While making arrangements to leave, he received another job offer from KFRC San Francisco.  Both involved the prestigious morning drive shift. Job offers required confidentiality while arrangements were made and in this case, confirmation of the preferred job at KFRC would have to wait until PD Sebastian Stone returned from a trip to London with the Rolling Stones. He was to return on a Monday, Beau's moving day. The phone didn't ring so Beau called KFRC. A nervous secretary informed him that Sebastian Stone 'doesn't work here anymore.' So he asked to speak to GM Hap Trout, only to learn he no longer worked there either (resulting directly from the billing irregularities at KCBQ). So he went to Chicago.

WEFM was in the process of changing formats from classical to Top 40 music. The night before the new format was to be kicked off, the 'Citizens Committee to Save Classical Music on WEFM' got an injunction to delay any broadcasting under the new format. Everything came to a stop while the judicial process intervened. It was at this time that Jack McCoy called to offer Beau weekends at KCBQ. 'Weekends? I can't live on weekends'. But his job options had quickly collapsed so he was off to San Diego.

During his first night on the air, his former PD, Michael Spears walked into the control room. He had driven down from Los Angeles after meeting with National PD for RKO Radio, Bill Drake, to tell Beau he had just been hired to replace Sebastian Stone as PD at KFRC and he wanted Beau to be his morning drive man. He soon received another job offer from KHJ PD Paul Drew to do mornings in Los Angeles. But owing to confidentiality, he could not disclose the job offer from KFRC. Drew was angry that Beau would not disclose anything and hung up on him. Off to San Francisco he went to join KFRC.

All of this occurred in 1973. There is an epilogue to this saga. Beau was at KFRC for three months when Paul Drew was appointed to replace Bill Drake as National PD of RKO. One of his first acts was to fire Beau from his morning drive shift. Spears managed to keep him on doing evenings for another year.

Gary Kelly (Weekends) replaced Beau Weaver about mid-year when he went to KFRC. Gary was a local that hung around the station to learn about radio. Eventually he got his broadcast license and worked swing shifts and weekends.

Jim Barker was a visible member of the office staff who wanted to be on the air, but lacked experience and education. He did have smarts, so he sailed through broadcast school exams, and got some limited air time. Jim was originally from Oxnard and knew Dave Conley when he was at KACY. Dave was probably instrumental behind Jim getting hired by KCBQ. Ultimately, his talent laid in other areas. He served as the station music director and tabulated music research to assemble the Top 100, Top 300, and other special edition publications. He was included in a DJ photo on the year end survey cover. He later became program director at KACY Oxnard in 1979.

Sam Diego (Steve Taylor) (Weekends) arrived from WHBQ in Memphis, but stayed only briefly before leaving to KFRC in San Francisco, where he used the name 'Citizen' Bill Carpenter.

1974

Mike Butts (6-9 AM) arrived from KIQQ in Los Angeles to do his Butts in the Morning show. Mike wrote for LA Radio that

"KCBQ . . . was created from the twisted mind of Buzz Bennett. The music was phenomenal. The Southern California energy was superb. I've said many times, when you push a button on the Q board and one of those shotgun jingles fired off, you'd better hold on for dear life and match the energy or you'd fall off." He later returned with Rich Brother Robbin to KIQQ. His last radio stint was in Providence, RI, where he now serves as the Deputy Director of the Mayor's Commission on Drug and Alcohol prevention.

Chuck 'Magic' Christian (6-9) performed the Drink to Safe Driving demo on his program by drinking on air demonstrating the cumulative effects of alcohol on his speech, linking this to its effects on a driver's judgement. He remained on staff through the year before being fired in December as Bartell Group PD Jerry Clifton moved to bring in staff he had worked with at WXLO.

Bill Moffitt (9-Noon)

'Shotgun' Tom Kelly (Noon-3PM) with an acoustical tone to his voice, Shotgun expanded his experience to include children's television work. He hosted the Emmy Award-winning *Word's-A-Poppin'* for five years on Channel 10 in San Diego and throughout the McGraw-Hill network. Later, he did more children's programming at KUSI TV in San Diego.

Dave Conley (3-6 PM) Conley moved on to become an original DJ at B-100 in 1975 and then KTNQ in Los Angeles. He held various PD jobs at radio stations in Texas. Rumors of his death in the summer of 2001 are apparently not true.

Gene Knight (6-9 PM)

Jimi Fox (Norbert Gomes) (9-Midnight) was hired by Roberts to do a night shift and serve as music director before jumping ship to B-100 (KFMB) in February 1975. He moved on to 10Q radio in Los Angeles then returned as program director. He now tends to his first love as a world class orchid grower in Manhattan Beach, CA.

Dick Young (Midnight – 6 AM)

Billy Martin (weekends and production) was hired to fill in for Dick Young on overnights, then continued to work weekends and perform production work. He earned a degree in Radio & TV from San Diego State University, (that included a class instructed by Desi Arnaz). Billy has also worked at KBBW, KDIG, KPRI, and KSON. He moved on to B-100 in 1975, then left radio. Billy has taught broadcasting courses at local colleges and worked weekends at several stations. He is now a successful psychotherapist in the Del Mar area.

Changes were occurring in the Bartell organization. Long time Group Program Director George Wilson was promoted to President of Bartell after hiring Chuck Roberts (John Dakins) from WXLO in New York as PD in August. Chuck was a solid broadcaster who possessed great production skills. He followed the basic Buzz Bennett format established in 1971. He had also been with several Bartell stations including KSLQ in St. Louis, MO Following his promotion, Wilson then hired Jerry Clifton, also from WXLO, to succeed him as Bartell Group PD. One of the changes Clifton insisted on the use of music directors. This would allow more time for the PD to handle production and promotion functions without the distraction of record label representatives pitching songs. Jerry Clifton now owns and operates Clifton Radio, Inc. based in San Diego.

1975

By this time, the station music selection had lost its cutting edge. Songs would not be added to the playlist until they reached Top 10 in local record sales. This helped set the stage for KFMB to move FM from beautiful music to a conventional Top 40 format as B-100 in March 1975. Roberts brought in former staff, including Brian White, from WXLO. The combination of Roberts as PD and White as music director were instrumental to improved music selection, using Grateful Dead/Jefferson Starship/Beautiful Day imaging. The $10,000 Visible Vault, I Q In My Car, and the

KCBQ RADIO AM 1170

annual Q-Walk for Mankind charity drive also took place during the year.

Jay Stone (6-10 AM) was hired from WXLO by Chuck Roberts to replace Mike Butts.

Bill Moffitt (10 AM-2PM)

'Shotgun' Tom Kelly (2 PM-6PM)

Brian White (6-9 PM) "The Blind Owl" was hired from WXLO by Chuck Roberts in January. He did the evening shift and served as music director after Jimi Fox left. Brian is credited for developing music with high appeal to the San Diego audience. He grew up in San Diego and had been on the air at KSEA in 1972 under Clifton. It was Clifton who brought him to WXLO, so their connections were pretty strong. He counts KCBQ as one of his best experiences in radio as a DJ. He is now Operations Manager for Cumulus Media, and also serves as PD and handles the afternoon drive shift at WKOO-FM in Jacksonville, NC.

Gene Knight (9 PM-1 AM) landed at B-100 in 1976 when PD Gerry Peterson (coincidentally similar to Gene's real name) wanted to bring his own people in. Gene was also at 91X, and ventured into the LA market at KHTZ before returning to B-100 in 1980. He had the foresight to start collecting radio surveys and other material during the mid 1970s. He has been a constant source for San Diego radio history and memorabilia in the 2000s. He now does the afternoon drive at KYXY in San Diego.

Dick Young (1–6 AM)

Tommy Sarmiento (Production and Week-ends) built a 24 year career in radio starting in the KCBQ mail room in 1971. His other duties involved circulating weekly

survey distribution, chauffeuring station staff, and researching music for air play. But Tommy seized on opportunities. He was hired by KGB in 1972 as a research assistant in the news department, then became the first overnight DJ at fledgling KSEA in 1972. That position led to a part-time job at KYA in San Francisco doing production and swing shifts that served to refine his skills. He was hired by Chuck Roberts and landed on the air the day following knee surgery due to staff shortages. But Tommy's forte' was in production work. He became known as the computer guy around the station. He followed a lead that KDEO was returning to a Top 40 format in 1976 under new owner Lee Bartell.

Chuck Geiger (Weekends) grew up in the San Diego area and had interned at KSEA in 1972 where Jerry Clifton was the station PD. Now that Clifton was the Group PD for Bartell, he suggested that Chuck Roberts hire him. Chuck arrived from KXFM in Santa Maria about October 1975 to work weekends. His air time increased as jocks came and went during 1976.

1976 – TOY SOLDIERS TO SINGING COWBOYS

Gerry Peterson (Gerald Cagle) arrived as PD in January when Chuck Roberts was transferred to KSLQ in St. Louis. Gerry had been consulting at KYA in San Francisco, and had been the PD at KHJ in 1974. But the station was moving away from a Top 40 format in favor of soft hits and KCBQ offered the Top 40 format he preferred. His arrival at the station was not immediately successful. He produced a new promotion called *KCBQ Becomes a Thing of the Past* to counter the rapid rise of B-100 on the FM dial. It was intended to promote KCBQ as an AM-stereo concept. But it was presented in the context of 'this is the station you grew up with, so please keep listening'.

'Black Tuesday' has been ascribed to the day that the staff was summoned to Jerry Peterson's office where he had toy soldiers standing on his desk. He told each jock which one of the soldiers was him and removed it, symbolizing that their tenures at the station had come to an end. One by one... Bill Moffitt... Gene Knight... Brian White... were released from their jobs. With the additional departures of Dick Young, Shotgun Tom Kelly, Jay Stone, and Chuck Geiger the turnover was complete. This did no diminish from the success Peterson would have at KCBQ, nor at KHJ that followed. But 'Black Tuesday' was the start of a series of events that would ultimately remove KCBQ from the Top 40 format.

The departing staff generally had long-standing ties with the community, some had actually grown up in the San Diego area, and all counted KCBQ as among the highlights of their careers. There was a connection between the staff, the station, and the community that was a continuum from previous staffs. But this asset had been devalued in the minds of current management. The incoming staff were no less competent and entertaining. Indeed, most have enjoyed highly successful careers in other markets and in syndication. But there was no sense of connection with the community nor does KCBQ appear to have ranked highly in their career ladder. A key reason may be that the outgoing staff enjoyed ratings leads that were essentially uncontested for several years. The incoming staff faced frustrations caused by the emergence of other strong stations in the market and the reality that FM was capturing the Top 40 audience. The shifts were adjusted to eliminate a position.

The line-up at the beginning of the year was as follows.

Jay Stone (6-9 AM) It has been demonstrated over and over that relationships are the strongest asset to protecting job security. Chuck Roberts offered that security, but he was now gone, and so was Jay when Gerry Peterson fired him in April. He went to KTLK in Denver. He was killed in an auto crash as he suffered a heart attack on October 15, 2001 in Honolulu, HA.

Brian Roberts (Jerred Persten) (10AM-2 PM) Brian got his start in radio while stationed in the Air Force at Duluth, MN. He worked weekends at station WEBC in Duluth on Sundays playing religious tapes and announcing the station calls. Following his discharge, he went to the Don Martin School of Broadcasting and earned a First Class License. His career took him to KYA in 1972. He was given the name Brian Roberts by KYA PD, (an former San Diego area DJ), Chris Cane. Chris had been at KCBQ and KGB earlier. Brian arrived in January with Jerry Peterson from KYA looking forward to a return to the Top 40 format. Brian is famous for his outspoken positions, and had made it known he felt the San Diego market was a step down from San Francisco. During a staff meeting in June, Peterson knocked another toy soldier over on his desk and told Brian that he was that soldier and would have to go. As a heated discussion ensued, Brian was released immediately.

'Shotgun' Tom Kelly (3-6 PM) has gone on to win the prestigious DJ of the Year in 1976 and won Jock of the Year for the Oldies category by Radio and Records in 2000. He still does oldies on KRTH radio in Los Angeles. He remains a radio legend in San Diego and resides in the El Cajon area. He left for B-100 where he would remain until 1980.

Mason Dixon (2-6 PM) arrived from WHBQ in Memphis to replace Shotgun Tom as the shifts were realigned to reduce staff. (Recent photo)

Domino Rippey (6-10 PM) arrived from WAPE in Jacksonville, FL to replace Gene Knight and stayed most of the year.

Dean Goss (2-6 AM) followed Dick Young in this shift in January.

Chuck Geiger (Weekends) had begun doing a lot of swing shifts as the station became short of staff in between the firings and hirings. Chuck was transferred to station KSLQ in St. Louis in July where he rejoined PD Chuck Roberts. Chuck Geiger is now PD and does afternoon drive at WLEV-FM in Allentown/Bethlehem, PA.

The lineup after June featured the following:

Charlie & Harrigan (Charlie Brown aka Jack Woods and Irving Harrigan aka Paul Menard aka Don Chapman) (5:30-10:00 AM) arrived from KFMB-AM where they had been from 1972 to 1976. The complete story of this team would require a separate book. The act dates back to the Murphy and Harrigan team formed in 1959. Charlie and  Harrigan was inaugurated in 1962 at KLIF in Dallas, TX. The team of Menard and Woods dates back to 1966. They remained popular at KCBQ from the Top 40 era through the Country and Western era spanning the years 1976 to 1984. They returned to the station in 1992 and 93.

Tony Maddox (10 AM -2 PM) acquired the nickname of 'Mad Dog' from Dean Goss that stuck with him.

Mason Dixon (2-6 PM)

Domino Rippey (6-10 PM)

Dean Goss (10 PM-2 AM) moved to this shift upon the arrival of Danny Wright. Dean's show featured an amateur minute where a listener could become the DJ at the bewitching minute of midnight each weekday night.

 **Danny Wright** (2-6 AM) arrived from Nampa, ID to work the overnight shift. (Recent photo)

Bob Shannon arrived from KWIZ in Orange County and stayed briefly before moving on to KFI. Bob left radio in 1983 to pursue a career in films and television. Now known as R.J. Adams, he has appeared in over 200 television dramas, as well as in the movies *Rocky* *IV* and *The Execution*. He now owns Shannon and Company, producing historical documentaries and films.

Ron Leonard (Weekends) 'The Boogieman' landed at KYTE in Portland, OR in early 1978. He was very popular with teens during the disco era.

1977

The station ratings plummeted in April as KDEO, now using the calls KMJC 'Magic 91', returned to a Top 40 format. The format featured a tight play list of heavily researched music. KMJC actually beat KCBQ in its first Arbitron ratings sweep. But the tight play list failed to hold an audience accustomed to a broad range of music. However, the damage had been done by making it easier for B-100 to draw listeners to the FM band.

Charlie & Harrigan (5:30-10 AM)

'Mad Dog' Tony Maddox (10 AM-2 PM)

Mason Dixon (2-6 PM) went to KHJ in Los Angeles by June. Mason later went to Tampa, FL where he remains active on the radio and in the community.

Steve Goddard (2-6 PM) attended Oklahoma State University where he broke in to radio while working at two college stations and at WKY. He is a prolific record collector and has built a collection that includes every hit from the top 100  list since 1955. He has converted that into a syndicated radio program called *Goddard's Gold* where he provides insider information about the song. (2003 Photo)

Jon Fox (6-10 PM) acquired the nickname 'Sweet Daddy'

Danny Wright (10 PM-2 AM) moved on to KROY in Sacramento. He now has a syndicated overnight show called Danny Wright All Night. He enjoys writing and has authored non-fiction books, short stories, and screen plays. He is also a prolific reader of any subject matter.

Dean Goss (2-6 AM)

1978

Mike Stafford, as GM, placed an emphasis on strong production. The station observed a continual drain of the adult audience to B-100. Jimi Fox began the year as PD. Meanwhile, Charter Broadcasting acquired the Bartell radio chain.

Joel Denver was programming Charter-owned WMJX-FM in Miami. They felt that the success he was having there should be applied to KCBQ in an effort to bring the 18-34 audience back. An aircheck of this year displays a considerably more mellow KCBQ than a few years before. The format offered a balance of current hits and oldies, intermixed with Hollywood updates and news from the legendary J. Paul Huddleston. Joel was on board only briefly as PD from August to January, but the target audience ratings went way up.

The line-up by August was as follows:

Charlie & Harrigan (5:30-10:00 AM)

'Mad Dog' Tony Maddox (10 AM-2 PM)

Steve Goddard (2-6 PM)

Jon Fox (6-10 PM)

Dean Goss (10 PM-2AM)

Linda Fox (2-5:30 AM)

Tony Evans (Jim Nelson) (Weekends and Production) began his career in 1967 in Phoenix, AZ and came to KCBQ from 10Q in Los Angeles where he had worked with Jimi Fox.

1979

Joel Denver, like Chuck Roberts before, was transferred to KSLQ in St. Louis as a reward for his programming efforts at KCBQ.

Jon Fox was elevated to program director while retaining his night shift.

Charlie & Harrigan (5:30-10 AM)

'Mad Dog' Tony Maddox (10 AM-2 PM)

Steve Goddard (2-6 PM) continues to syndicate *Goddard's Gold* and is presently with WELW Willoughby, OH

Jerry G. Bishop (Jerry Ghan) (2-6 PM) was a very busy man in San Diego when he started at the station in 1979. He was already hosting Sun Up San Diego, a live daily talk show on KFMB-TV, and would continue to do so until 1990. His career began when he sang in a folk trio during college and applied for a job at WNMP in Evanston, IL for an income of $67.50/week. He moved on to KYW in Cleveland where his air name was altered to Jerry G. When he was hired by station WCFL he was asked to add a full last name, and Bishop emerged from scrutinizing phone directory listings. His friendship with Charlie and Harrigan led to an offer to come to KCBQ. Jerry notes, "I was not thinking about how tired I'd be every night unless I took a nap between Channel 8 and Q . . . most days I didn't." He left in July 1980 when the first of his two restaurants opened as Seaport Village, and as the station was moving in to a Country format that didn't interest him. He remains on the air at KPOP (formerly KGB), and remains busy with his two restaurants.

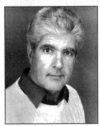

Jon Fox (6-10 PM)

Perry Allen had previously been at KDEO during the oldies format era.

Linda Fox (2-5:30 AM)

Tony Evans (Weekends and Production) returned to the Phoenix market where he does afternoon drive at jazz station KYOT and production work at KESZ-KFYI and KGME.

KCBQ RADIO AM 1170

1980

The Top 40 format at KCBQ came to an end in 1980. Country and Western formats had made substantial gains in markets across the nation on the AM band and KCBQ followed that trend. Ironically, some names reaching deep into the station's past re-emerged as Country DJs. These included Harry Scarborough, Bill Moffitt, and Charlie and Harrigan. Phil Flowers returned as a traffic reporter and to do production work. The station has continued to follow national trends by simulcasting AM and FM and evolving through Country and Western, Oldies, and to its current all talk format.

The demise of KCBQ can be traced to 'Black Tuesday' when so many jocks with long-standing ties to the community were fired. This was followed by market gains enjoyed by KMJC, and then the mainstreaming of Top 40 on the FM band that brought huge ratings to B-100. Another factor had to be the frequent turnover of ownerships, leaving no sense of how the station would position itself in the community or in the market.

1990

After many format evolutions into the synthesized product now heard on radio today, KCBQ hosted and aired an all day DJ reunion on March 1, 1990 at the Marriott Hotel in Mission Valley. Disc jockeys who had worked at the station extending back to 1955 had one more opportunity to say the KCBQ call letters on the air while reminiscing about their experiences and introducing oldies music. Many of the people who participated, and whom we grew up with, have since died. The history recorded during this event was invaluable.

2003

The site of the transmitter and former studio at 9416 Mission Gorge Rd. in Santee was sold to make way for a Lowe's store. KCBQ maintained a lease for the transmitter and towers, and Jack McCoy's company leased the studio, through September 2003. KCBQ analyzed 27 alternate locations to move the transmitter and towers, and selected a property near Muth Valley Rd. in Lakeside. After gaining FCC and FAA approvals, an application was filed with the County of San Diego. Opposition from the community, and County staff led to a denial by the Board of Supervisors on April 9, 2003. The station faces either temporary, if not permanent, removal from the airwaves; or sharing space on the transmitter of KPOP, (formerly known as KGB). The shared tower option would require substantial reduction in power transmission and bring the 50,000 watt era to a close

---

FOOTNOTES

[1] *The Hits Just Keep On Coming,* Ben Fong Torres, 1998, P 155

[2] *Dream-House,* Bill Earl, 1991 Second Edition (Red cover)

[3] *The Reunion 1955-1990, KCBQ Radio,* California Aircheck, 1990

[4] Personal daily diary of Bill Earl, 1965-196

Images – KCBQ Radio AM 1170

Match cover from early 1950s.

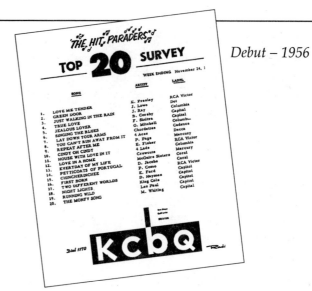

Debut – 1956

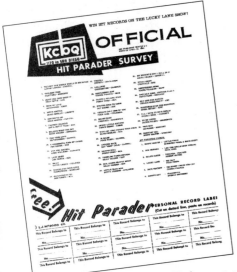

Debut – March 1958

Debut – August 1958

Debut – 1960

Debut – August 1960

Debut 1–961

Survey – 1962

Different Banners – 1963

Bill Bishop, 7th & Ash Studio – 1963
Note the glass window to the street, the mirrored ceiling, and Casey B. Quack on the floor.

Qoodle Pad

Artist covers were used during 1964.

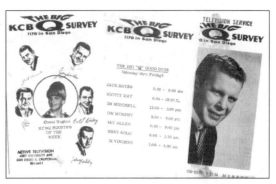
A variety of forms appeared in 1965. The DJ album started, followed by individual DJs, program schedules and advertising.

Austere surveys during 1966 to Summer 1967 featured a DJ, Artist or Ad.

GO Magazine included the weekly survey from summer 1967 to March 1969. The 1968 and 1969 editions of GO offer great insights into station operations and personalities.

1965 – Beatles Concert

KCBQ RADIO AM 1170

Side and front views of Santee Studios in 2003.

Debut – April 1969

KCBQ Transmitter – 1966

Debut – November 1969   *Debut – April 1970*   *Debut – November 1970*

KCBQ RADIO AM 1170

Debut – February 1971

Debut – 1972

Country & Western Debut – 1981

Debut – 1973

KGB Radio AM 1360

Inaugural Top 40 Format

When Brown Broadcasting bought ABC affiliate KGB radio in 1961, they elected to stay out of the daily operations of the station and focus more on strong program directors to generate success. Not even the General Manager was a factor in what went on the air. This philosophy made KGB a format-driven station. This story identifies 69 air personalities working under eight program directors over 19 years of the Brown Broadcasting ownership.

1963

Prior to the Drake and Chenault years, KGB's MOR format included Lawrence Welk, George Hamilton IV, Steve Lawrence, Nat King Cole, and other 'heavy rockers' on their play list. By fall 1963, the play list featured a solid rock n roll line-up, but the station was still not formatted to succeed. The station continued to use an ABC news network feed, but listeners interested in extended news coverage would not be tuned in to a Top 40 station. The Top 40 audience preferred the music-oriented formats of KCBQ and KDEO. The original *Silver Dollar Survey* was on an indistinguishable 8 1/2 X 11 sheet with the slogan *You'll like the New KGB*. By the end of 1963, a more stylized bi-fold *Silver Dollar Survey* was published weekly introducing photos of the DJs as the *Station of the Stars* with a 40 song play list that featured artists with more appeal to a younger audience. The station was moving in the right direction under PD Dick Drury, but not enough to capture substantial ratings. The DJ line-up featured some recognizable names to local listeners before their KGB employment.

Art Way was formerly with The Mighty 690 XEAK doing taped shows at the Mission Valley Inn for broadcast the following day. He probably came from KDWB Minneapolis where he was on staff in 1961. After leaving KGB, Art had an afternoon show at KOGO Radio in 1969 where he was reunited former XEAK colleague Ernie Meyers.

Bill Wade (Bill Wade) was formerly with KDEO radio in San Diego and would remain with the station through numerous format and staff changes, anchoring the Noon – 3PM shift. He did a brief stint on weekends at KHJ Los Angeles in 1963. It was there that owner Willet Brown met him and referred him to Dick Drury to come to KGB within a few  months. Bill was on the air when Bill Drake came to town. Drake invited Bill out to dinner, and while nursing martinis, Bill told Drake he need not buy him dinner to explain he was getting the axe. Drake replied, "what are you talking about? I invited you to dinner to ask whether you wanted the Noon to 4 shift or the 4-8 shift?" Bill retained his Noon to 4 shift and the two remained friends over their careers.

Dick Drury was not known in the area before coming from WIL St. Louis in 1962. His career  began at age 15 in 1950 at WSRS Cleveland, and moved into the dual role of DJ/Program Director at KISN Portland, OR, then at KQV Pittsburg in January 1960. He also did the morning drive shift and served as Program Director of KGB beginning in 1962. He left the station just prior to the arrival of Drake. He continued to have successes at KLOS Los Angeles, KRQK Lompoc, and promo work at KHJ during the 1970s. By 1979 he was Director of Programming for the Susquehanna Broadcasting network in the Pittsburgh area, then moved into station ownership. He has since died.

King Richard (Richard Widenbrook) was formerly of KWK in St. Louis. He had originated listener 'raspberrys' that sounded like loud gas. He landed in the hospital after responding to a girlfriend's marriage inquiry with laughter. He had already made plans to leave the station in early 1964.

Other staff included **Bill Masey** (below) and **Jim Mitchell.** Mitchell was on staff in fall 1963 for three months. He later appeared at KCBQ in 1965, and on KHJ news in 1966.

Boss Radio
Phase One: 1964 Getting Started

With Gene Chenault doing management consulting, and Bill Drake doing format development, the team revolutionized the pacing format of Top 40 radio that improved ratings. They began their association with KGB in April 1964 and had their format and staff in place by summer. They expanded quickly to KYNO Fresno and KSTN Stockton. KHJ Los Angeles was added in April 1965, followed by KFRC in February 1966. KHJ was considered the flagship of this unique program network. The term 'network' had denoted common-held ownerships up to this time. The 'Drake' network denoted common format stations with different owners. Drake's three big California stations that he programmed were KFRC, KHJ, and KGB.

The format featured more music each hour, as automation replaced manual signals between the DJ and the engineer. The following rules applied:
- 14 songs were to be played per hour
- DJ chatter was limited to 15 seconds between songs
- DJs were to introduce one song as the prior song faded out
- Station jingles were limited to 90 seconds
- Advertising spots were limited to 13 minutes
- News reports occurred 20 minutes past the hour rather than at the hour
- Songs were played back to back to avoid listener dial switching

Drake described his format as, "Everything tight, no dead air, . . . everything perfect, perfect, perfect." In doing so, DJs were easily interchangeable among the stations they programmed. Seven DJs are recognized for this feat.

Under the auspices of the Drake and Chenault consultancy the entire staff was overhauled, leaving only DJ Bill Wade to remain. John Wrath was General Manager. Fred Lewis did the news aided by Ted Marvelle who was news editor and helped on promotions. Among the promotions was the Big Kahuna, (a Samoan who danced at Mission Beach resorts with his family and the big Woody that traveled along area beaches handing out prizes to promote the station). The station also experimented with a Beach Boys format to reflect San Diego as a Beach City.

The station published the *KGBeach Boys Top 30 Survey* that introduced the slimmed down play list. All of the DJs, featured as KGBeach Boys surfers, were pictured on the front of the weekly survey. The station published the Top 100 songs on 1964, and began making gains on KCBQ and KDEO, but did not emerge as the ratings leader. Talent had to match the investment and KGB was loaded.

Tom Maule (Tom Maule) (6-9AM) led off the day, coming from KMAK in Fresno. He would remain with the station until 1967, when he moved to 93 KHJ in Los Angeles, then KFRC in San Francisco. Standing at 5 ft. 6 inches, Tom was noted for his mix of humor cynicism and for being caustic at times, but always a great talent. He was actually a competitor to Chenault's KYNO at Fresno. He is only one of seven DJs who did service at the three major Drake-Chenault stations in California. Tom died in 1993.

Les Turpin (9-Noon) arrived from Chenault's KYNO in Fresno as Program Director and did a mid-

morning show he called 'Turpin Time'. He pulled himself off the air in 1965 to focus on PD duties, but later returned to the morning shift. He remained until 1967 when he went to work for the Drake organization, then to KFRC in San Francisco.

Bill Wade (Noon-4PM) continued to reign over the afternoon drive shift despite difficult work relations with PD Les Turpin.

Steve Jay (Steve Janovick) (4-8PM) has been on the air in Southern California for nearly all his career. That career began at KFIL-FM in Santa Ana in October 1962. This led to stations in Victorville, Palmdale, and San Bernardino before moving upstate to KMAK in Fresno. He arrived in April from KMAK and held the afternoon drive shift. He switched to the overnight shift, with Johnny Hayes moving into this shift, by the end of the year. He resumed the afternoon drive shift in April 1965 when Ray Morgan arrived to do overnights.

Chuck Cooper (8-Midnight) also came from KMAK in Fresno and did the evening shift. Chuck had been fired at KMAK and was offered a job by Bill Drake in Stockton. But he lacked the Class I license that the job required. So he  enrolled at Ogden's School of Broadcasting at Santa Barbara and earned the license. He then set out looking for a job and made a cold-call visit to the KGB studio with a tape. He did not realize that Drake was now running the station. He left the tape and received a call from Drake that night who hired him to begin the next day. Chuck left in January 1966. He was on staff at KCBQ in 1976. Chuck is still in the business doing production work at KOGO in San Diego.

Johnny Hayes (John Hayes) (Midnight-6AM) began his career in his hometown of Macon, GA at WNEX in 1959. While traveling through the area, Bill Drake heard Johnny and offered him a job at Bartell-owned WAKE in Atlanta. Within a year, he moved to sister station WYDE in Birmingham as Danny Day, then on to KYA in San Francisco. Drake hired Johnny to become an original *KGBeach Boy* in 1964. Johnny began as the overnight DJ, then switched shifts with Steve Jay by the end of 1964.

In May 1965, not knowing that KHJ in Los Angeles was planning to make him an offer in their original *Boss Radio* lineup, Johnny accepted an offer to move to LA's top rated station at the time, KRLA. The move was initiated by KRLA DJ Emperor Bob Hudson, with whom Johnny had worked at KYA. Except for portions of 1968, 69, and 71, Johnny remained with KRLA for 27 years. While there, he created the 'Big 11 Countdown Show' that he hosted for more than a decade.

Johnny's most recent move came in 1992, when again, Bill Drake, then consultant to KRTH in Los Angeles, heavily promoted Hayes. Johnny remained there for 10 years, until June 2002, and now lives in semi-retirement in the Hollywood hills. During his 43 year career, Johnny was named "Oldies DJ of the Year" for two consecutive years by Billboard Magazine, and earned a star on the Hollywood Walk of Fame.

Ray Morgan (Midnight-6AM) arrived from KMEN San Bernardino by April 1965 replacing Steve Jay on the overnight shift after Steve switched shifts with Johnny Hayes.

Jerry Davis (Jerry Davis) (Weekends) was attending college while interning and working for wages at KFMB-TV, KPRI, and KGB. Jerry worked at KGB from June to November 1964. He was originally hired to edit news copy and do phone interviews with newsmakers for airing. Jerry eventually got on the air for about six weeks on Sunday nights and the post-church shift on Sunday mornings. He joined AFTRA when he went on the air and was paid the union scale at $3.63 an hour. Jerry eventually moved from radio to become a video technician for CBS-TV. He edited most CBS comedies from 1972 (*All In the Family*) to 1999 (*Everybody Loves Raymond*). "It was a pretty wild ride, but that early sixties part still holds memories I'm fondest of."

BOSS RADIO
Phase Two: 1965 - September 1966

By Fall 1965, KGB was No. 1 in the San Diego market under the Drake-Chenault format. The weekly survey showed the transition underway by being retitled the *KGBeach Boys 'Boss 30'* with all the Boss Jocks on the front.

Gary Mitchell joined the staff briefly at the 9-noon shift replacing Les Turpin by September 1965. Les had taken himself off the air to focus on PD duties.

Bob Elliott (Bob Early) arrived in February 1966

from KXOA in Sacramento to replace Gary Mitchell in the 9-Noon shift. Elliott had been on the staff of KYNO in Fresno under Chenault before moving on in 1965 to KEWB in Oakland, then to KXOA. Bob remained through October 1967 when he left for an east coast job. Bob would resurface at rival KCBQ in 1970 under the name of K.O. Bayley. He returned to KGB in 1971 as K.O. Bayley. Prior to becoming a DJ, Elliott, at 6'3", had been a professional baseball player and a professional boxer suffering a broken nose in the ring. He never drove a car, possibly due to poor eyesight, and was always shuttled to and from work by his girlfriend Pam.

A series of line-up changes began in January 1966 with the departure of Chuck Cooper and Ray Morgan. The new line-up was as follows:

Stan Walker (Jim Scott) (6-9 AM) arrived from WENE in Endicott, NY in January to replace Tom Maule in this shift. This was the only west coast station that Jim worked at in his career, and the only station where he was given the name Stan Walker, to

avoid confusion with Scotty Day of KCBQ. "I had a good time there and learned a lot about a tight streamlined format, but it wasn't really the type of radio I was interested in." Jim does the morning drive show at WLW Cincinnatti and was winner of the 2002 Marconi Award.

Bob Elliott (9 AM-Noon) continued in this shift.

Bill Wade (Noon-4 PM)

Steve Jay (4-8 PM)

Tom Maule (8 PM-Midnight) took this shift vacated by Chuck Cooper.

Jim Meeker (Midnight-6 AM) arrived in January to replace Ray Morgan but was disenchanted with the overnight shift and left for WPOP Hartford, CT in April.

This caused the following adjustments to the line-up:

Les Turpin (6-9 AM) had to resume a shift because the station was short of staff, and opted for the one held by the newest DJ, Stan Walker.

Bob Elliott (9 AM-Noon)

Bill Wade (Noon-4 PM)

Steve Jay (4-8 PM) was completing his tenure with KGB as Drake offered him a job at KFRC in San Francisco. After his KGB years, Steve Jay became known as Jay Stevens. This was necessary to distinguish him from Steve O'Shea who was already at KFRC when he arrived. He did weekends at KRTH Los Angeles as Steve Jay until his retirement in 2003.

Tom Maule (8 PM-Midnight)

Stan Walker (Midnight- 6 AM) was moved to this shift to replace Jim Meeker.

By September, the Beach Boys format was eliminated from the station marketing plan in favor of a Drake network unified theme around Boss Radio. DJs were now *Boss Jocks*. The weekly survey was now the *Boss 30*. The Boss theme and format were enormously successful. The station published the Top 100 hits of 1965 that featured a drawing of the KGB studio on the cover.

BOSS RADIO
Phase Three: October 1966 - May 1969

The Chenault-Drake format continued to develop during 1966. A revamped *Boss 30* survey was begun on October 26, 1966 with Official Issue No. 1. This format was pocket size and tracked the weeks a song was on the list in addition to current and past week positions. A photo of one disc jockey or contest prize was featured each week.

Mark Denis (Denis Melbourne) joined the staff by October 1966 in the 6-9AM shift from KMEN in San Bernardino. Mark would remain  until 1969 when he went to KFI in Los Angeles. He was well known for his work as a traffic reporter at KFI over the decades. The *Mark Denis Melbourne Interchange* at Highways 91 and 55 in Orange Co. CA was named in his honor in 2002. Mark was famous among his friends and acquaintances for calling them to offer his best wishes on their birthday. He died of a heart attack in 2000.

Les Turpin moved to the 9-12 AM shift and Bob Elliott took the 4-8 PM shift when Steve Jay left to KFRC in San Francisco.

Bill Brown (Midnight-6AM) arrived in August  from the Far East Network based in Japan to fill the overnight shift, replacing Stan Walker who went to WSBA in York, PA. He has been with WCBS-FM New York since 1969 and hosts the popular "Brown Bag Special" during his mid day show on Oldies Radio WCBS-FM. Bill has also provided English dubbing for Japanese films.

The station published a survey of the Top 100 hits of 1966 at the end of the year.

1967

The station increased their live concert promotions with top talents of the day during 1967. KGB produced its first album of *21 Boss Goldens Volume 1* in June. The album cover featured a cartoon of a geologic KGB formation rising from a crowd of hip people on the front, and drawings of the Boss jocks on the back. These same drawings were used on weekly surveys. The year also saw a substantial number of changes in the staff as talent was developed and moved on to larger markets. But, KGB owned the town from a ratings perspective, and may have been the year of its peak dominance in the market. The lineup had changed radically since the original 1964 Drake team and now featured the following by the end of 1967:

Jim Carson (Vic Gruppi) (Midnight-6AM) was known as Vic Gee at KBLA in Los Angeles before it switched formats on June 17th to become Country station KBBQ. Then he became Vic Grayson. He lined up a job at KGB in June. However, with military service looming in the summer, Les Turpin did not want to start him on the air then hire

KGB RADIO AM 1360

a replacement for a short time. So his on-air debut was set for September 1. When he arrived, Les Turpin had moved on to Drake network duties and Mark Denis was the new Program Director. He assumed the overnight shift vacated by Bill Brown when he went to WOR New York. He moved to the 8-12 PM shift in October when Dick Saint assumed the 4-8 PM shift when Bob Elliot left.

Jim brought a new image of informality to the station. Prior to his arrival, jocks usually appeared in formal coat and tie. Carson sported t-shirts that all his (new) colleagues would adopt in 1969. He remained with the station through April 1970 when he left to KFRC in San Francisco. Jim was a fill in host at KHJ for one week in 1973 that made him one of only seven to serve at the three big Drake-Chenault stations of KFRC, KHJ, and KGB. He is still on the air at oldies KRTH.

Mark Denis (6-9-AM), moved into the 6-9AM shift and became Program Director in July when Les Turpin left to work for the Drake organization. Les became program director at KFRC in November. As a PD, Mark's strengths were in production quality, leaving the majority of the programming decisions to the Drake consultancy. Mark also encouraged camaraderie among the broadcasters in town, going so far as to exchange gifts during the holiday season.

Bill Wade (Noon-4PM), the only original Boss jock was completing his last year with the station before being transferred to KFRC in San Francisco. He was reunited with his former, and then PD Les Turpin, so he left that situation as soon as he could. He arrived at KHJ before the end of the year. He later founded the Bill Wade School of Broadcasting. He has been an instructor at Arizona colleges, operates a firm with his family that designs and sells swimming pools, and has a ranch just outside of Phoenix.

Bob Elliott (4-8 PM) held the late afternoon shift until he left in October.

Dick Saint (Dick Middleton) (8-12 PM) arrived in June from KISN (Portland, OR) to take the night shift vacated by Tom Maule who went to KHJ in Los Angeles. Dick had worked with *The Real* Don Steele in Oregon, who took credit for christening him Saint during that time. He was known as a high energy exciting radio personality whose talents were best suited to the 4-8 PM shift when Bob Elliott left in October. He was one of the seven to work at all big Drake-Chenault stations.

Johnnie Darin (John Christian Miller) (9-Noon) arrived from KMEN San Bernardino in July when Les Turpin went to KFRC San Francisco. John dropped his last name early in his career; but then encountered Chuck Christensen when he arrived at KMEN. Seeing a flyer posted on the studio wall advertising a James Darren concert, he quickly adopted a modified version as Johnnie Darin. Darin hoped to eventually end up at 93 KHJ via the Drake network. He now is retired.

Dave Stone (Midnight-6 AM) came in October when Jim Carson moved to the 8-12 PM shift.

Ironically, by the end of the year, former KGB staff Les Turpin, Jay Stevens, and Tom Maule were all at KFRC in San Francisco, along with former KCBQ jock Johnny Holiday (under the name of Sebastian Stone). Dick Saint and K. O. Bayley joined the staff the following year. Turpin was Program Director at KFRC.

KGB published a survey of the Top 100 hits of 1967 at the end of the year that featured the autographs of the DJs on the back.

1968

The staff additions this year provided strong radio personalities who would remain popular in the San Diego market and very prominent in the Top 40 radio industry. But success did not come immediately. The introduction of all new afternoon and evening DJs, coupled with innovative programming at KCBQ, knocked KGB from the top of the ratings for the first time since 1965. At the same time, Drake tilted the music list toward pop and cross-over artists to avoid the more edgy 'progressive rock' cuts that were emerging on FM stations. This was intended to attract the adult audience and would continue until Buzz Bennett became PD. The *Boss 30* was redesigned in April from the Drake network format to a design unique to station KGB. The *21 Boss Goldens Vol. 2* was also released at this time. The album featured 'mod' flower designs above sketches of the DJs that were used throughout the year on surveys.

The situation was growing tense for KGB. When KCBQ PD Mike Scott adjusted his programming to place a music sweep against KGB's news he received a terse note from Mark Denis that read, "When are we going to stop this crap."

The shifts were realigned a follows:

Jim Carson (6-9 AM) took this shift when Mark Denis moved to the 9-12 AM shift.

Mark Denis (9-Noon) took this shift when Dick Saint went to KFRC San Francisco, and also served as Program Director during the year. Mark enjoyed radio production aspects the most and was a perfectionist at it.

Johnnie Darin (Noon-3 PM) took this shift when Bill Wade went to KHJ Los Angeles in May. He would leave in November and join KRLA in Los Angeles in December. He working initially behind the scenes in production, then taking the market by storm as DJ and Program Director.

Gene West (Tom Leland) (Noon-3) arrived from KROY in Sacramento in November. Tom took an interest in radio while serving in the Army. He shared quarters with a guy proclaiming to be a DJ, and Tom felt if this guy could do it, so could he. He soon landed a job at an FM station in San Francisco that eventually led to KACY in Oxnard, then to KROY. Tom developed his own air name of Gene West, in part as a tribute to his favorite DJ while growing up in Marin County, Gene Price; and from gazing at a weather vane atop his Mother's home and opting for 'west'.

The Drake network was *the* place to be during the late 1960s and Gene used his acquaintance with Johnnie Darin to pass along tapes for the Drake people to hear. Soon, he was offered a job at KGB, ironically replacing Johnnie Darin when he concluded that KGB was not his ticket to KHJ and went to help resurrect KRLA. Gene remained a rock of stability for KGB through many staff and format changes. Gene was the station music director, although all new music selections had to be cleared with Betty Brenneman who was Bill Drake's Music Director.

Dick Saint (3-7 PM) remained until October when he went to KFRC San Francisco. Later, he was with KHJ Los Angeles making him one of six who were on the air at all three Drake-Chenault stations. He followed Johnnie Darin as Program Director of KRLA in 1971.

Bwana Johnny (Rick Johnson)(3-7PM) began his career at KLOG in Kelso, WA. The name Bwana Johnny originated in West Palm Beach, FL. He was an original DJ at KISN in Portland, OR when it adopted the Top 40 format, and was on the air at KISN during its last broadcast in 1976. He arrived at KGB from WUBE Cincinnatti in late 1968 to take the afternoon shift. He did not like the area and stayed only a few months before leaving to KJR in Seattle. He was at KYA San

KGB RADIO AM 1360

Francisco in 1969, and now resides in Seattle. He is a program director and an air personality with the Jones Broadcasting Network.

Bobby Ocean (Ernest Raymond Lenhart) (7-12PM) Lenhart delivered tapes of himself to KFRC, KYNO, and KGB not realizing that these were all Drake-Chenault station. After Drake received tapes from the program directors, he was hired at KYNO and given the name Johnny Scott. He received a call from Bill Drake in May 1968 who asked if he wanted to go to San Diego. Johnny replied, "Oh, yes sir."

"You start Monday" Drake replied "..... and your name will be Bobby Ocean." His heart sank, but he knew better than to argue. He knew too many people who would go. [1] "I made a conscientious choice to be a pro when I arrived in San Diego." He replaced Jim Carson who moved to the 6-9 AM shift.

Bobby would become the signature voice and prominent personality for whatever station he was at. 'I'm always bill boarding, " Bobby later explained. "Coming up, something's always coming up. The curtain's always rising.' Ocean often repeated the Shakespearian line, "The love you take is equal to the love you make" as was later recorded in a Beatles song. He also is a cartoon artist and designed many covers and art work for contests in the weekly Boss 30. Bobby Ocean is the only person to have worked for all four Drake-Chenault stations.

Dave Stone (Midnight-6 AM). Held the shift through the year.

Don Dale (Don Eller) (Weekends) arrived in July to do weekends.

1969

San Diego radio listeners had the luxury of having two great top 40 stations with talented staffs. This created a very competitive environment between the local stations. On the matter of associating with DJs from other stations, Rich 'Everybody's Brother' Robbin writes that in those days, "I would have rather shoved needles through my eyeballs than associate with THOSE FUCKERS. They were the ENEMY. You need to understand back then we would have rather simply run them over or shoot them, such was the rivalry, unlike today where people daily commit the treasonist act of associating with the opposition. We would literally have eaten worms rather than done that."

KGB continued to develop its format to appeal to its target market. The number of shifts were increased to add more personalities.

The following line-up emerged by mid year:

'Mighty' Jim Carson (6-9 AM) was now senior among the staff. He served as interim Program Director after Mark Denis left.

Gene West (9-Noon) by his name presented rival KCBQ with a problem. They had sold ad spots for Jeans West, a local clothing store. KCBQ did not want it to appear that they were having their jocks plug a competitor as they read the ad, so they concluded with "Funny name for a man, great place to shop."

Gene was enjoying the celebrity status afforded to DJs in San Diego at this time. He was stopped by police while driving his red Dodge Charger as he was drunk. The cops were ready to throw the book at him until he gave them his air name. They were always fans and would let him go.

Christopher Cane (Kenneth R. Witt) (Noon-3PM) arrived in February to fill the shift vacated by Gene West, who was doing the 9-Noon shift following the departure of Mark Denis to KEZY in Anaheim. Chris and Gene would

switch shifts by August. Chris could often be found on the roof of the studio working on his tan, or at any rodeo event in town. He was known for his wit and was not as boisterous as many of his colleagues. Likewise, he was an announcer on his show, not an entertainer. Chris followed Buzz Bennett to KCBQ in 1971. He later returned to the station in 1979 as 13K after serving as PD at KYA in San Francisco.

Bobby Ocean (3-8 PM) moved to this shift following the departure of Bwana Johnny. Bobby was an advocate of the book *The Power of the Subconscious Mind* by Dr. Joseph Murphy and has attempted to practice and preach the elements advanced by Dr. Murphy.

Perhaps intending to explore his subconscious mind, Ocean was arrested for possession of marijuana. The case was thrown out due to an illegal search. The next weekend, Ocean was set to introduce Led Zeppelin at a local concert. Unsure of his standing in the community over this incident, Ocean walked out on the stage and said, "I'm Bobby Ocean of the San Diego Narcotics Squad." The crowd went wild with enthusiasm. That was when he knew he had 'arrived' and that his bond with the community was established.

Bwana Johnny (8 PM-Midnight)

Dave Stone (Midnight-6 AM)

Don Dale (Weekends)

BUZZ RADIO
May 1969 - August 1970

Until this time, program director duties went to the most senior disc jockey at the station. Since the Drake years began, Les Turpin, Mark Denis, and briefly Jim Carson, had served as PDs under the close supervision of the Drake consultancy. That was now going to change. Buzz Bennett was named national Program Director of the Year for 1968 after turning WTIX New Orleans around. Given the opportunity to bring him to San Diego, the Browns broke with tradition and hired a non disc jockey to program their station. (Photo from 1990s)

Buzz Bennett would have the effect of changing stations he programmed to national trend-setters. He started by doing extensive market research, including interviews with record buyers around town. From this, they identified 'power oldies' that listeners wanted to hear. New *Hit Bound* songs would follow the news. Long music sweeps were played while the competition was at their news break. The station jingles were set just prior to music, never as an intro to a speaking moment. The immediate result was a return to the top of the market ratings under his programming.

In addition to tight formats, he championed aggressive contests. Having fallen behind KCBQ the previous year when they offered larger cash jackpots, the Browns authorized Bennett to double the amounts KCBQ was offering as the *KGB Double Cash Jackpot*.

Bennett also had an ear for music and with the right song, he would program around it, independently of the control the Drake network normally asserted over music selection.

It is important to note that as Buzz settled into the Program Directorship, he took control of the station away from the Drake consultancy. The Drake format generally eliminated songs after eight weeks on the charts regardless of their continuing listener appeal. Buzz carried songs on the charts until public interest faded, or even longer if there was a programming advantage to be achieved. The Drake format usually featured air staff in an equal rotation on the *Boss 30* cover each week. Given the demographics of the young listening audience, Buzz opted to promote his key staff in the morning, afternoon drive, and evening shifts substantially more often than other DJs to further bond the audience greater with these personalities.

In addition, Buzz wanted no negativism on the air. All statements had to be positive, upbeat, and happy. Gene West came to work one day with a lousy cold and said so on the air. Realizing the mistake, he quickly followed it by saying it was the best cold he ever had.

Buzz was always thinking about what, if anything, could be improved, but was shrewed enough not to tinker with successes. The metaphor applied to Buzz is "The price of a clean window is tireless wiping". Rich Brother Robbin writes that, "Buzz's use of hallucinogenic drugs was part of a quest to bring something higher and more significant to what we did on the **radio**…better ways of communicating, of sharing, and lifting people." Buzz hyped *I'm OK, You're OK* by Dr. Thomas Murphy, *Godfather* by Mario Puzzo, and *Future Shock* by Alvin Toffler to the staff. He also brought about a different culture in the studio. Competitors were the enemy and their job was to (professionally) kill the enemy.

They listened to the competition, as well as each other, to identify ways to improve or grab an edge. The staff covertly rummaged the trash bins of their competitors to gather internal memos and other programming intelligence. The culture around the studio was also a drug culture not unlike the broader societal changes that were underway. The 'extended' studio was *The Animal Bar*, a bowling alley, luncheonette, and the nearest bar to the studio that offered all that the staff would want, including after-hours access. The staff became closely knit and tough. All they thought about at the time was radio, what they were doing, and how they could improve upon the product. Station owner Willett Brown's reaction to this phenomenon was to declared Buzz to be "his baby".

Rich Brother Robbin (Richard Werges) (8-12PM) The shift had opened up when Bobby Ocean took the 3-8 PM shift. Rich worked early in his career under names such as Rock Robbin, Rich Robbin, and Richard W. Robbin. By the time Rich arrived from KRUX Phoenix in May, he was using the nickname of Rich "Mutha" Robbin (after Gary "Mutha" Widdom of Union Gap). Buzz Bennett loved the nickname, but Drake said it was 'too coarse'. The night before he started, May 30, 1969, they went out to dinner at Chinaland Restaurant. Through the haze of saki, reefer, and unfiltered Camel cigarettes, Buzz said, "Well Ritchie, what're we gonna do about your name?" Out of nowhere Rich replied, "Well hell, why don't I just be Rich BROTHER Robbin and be everybody's brother?" The dinner party erupted into applause as Buzz Bennett tapped Rich once on each shoulder with chopsticks and said, "I christen you Rich Brother Robbin."

Don Dale (Don Eller) (Midnight-6AM) took the overnight shift in August when Dave Stone went to KYA in San Francisco. Don worked for the local telephone company after leaving radio, but has returned to coordinate live shows at KCBQ.

Bob Foster (9-12 PM) arrived from WIBG in Philadelphia in August to take a new shift. Foss, as he was known, was a huge man with a booming voice who was considered a bully by some colleagues and a teddy bear  by others. His focus was usually on his girlfriend. One night at *The Animal Bar* he got drunk, then mad at the manager. He decided to start bowling and heaved the ball so that it would crash into the pinsweeper door, thus causing the manager to demand that he leave the premises. He developed the Midnight Cowboy slogan at the time the movie was so popular.

Tony Richland (Record Promoter) was known to everyone in radio. As an independent record promoter, he would appear at the station with an arm full of records at least twice a month. As a friend of Buzz, he warrants recognition for having his photo spliced into several surveys in the late 1969- early 1970 period and leaving so many of us wondering who the hell this guy was.

1970

Boss Radio, the Buzz Bennett way, was big, making KGB the ratings leader for most of 1970. Only a few staff changes occurred as staff moved on to larger markets. The station published a survey of the *Top 100 of the 60s* at the beginning of the year that featured a cover of a Bobby Ocean drawing of a hand forming a peace symbol with photos of the DJs on the fingers.

Jesse Bullet (Richard Bullen) arrived from KACY Oxnard to do weekend shifts and stayed through the end of the Top 40 run in 1972. He was the only "Boss Jock" to remain on staff during the album progressive rock era that followed. Gene West, who knew him from KACY, had referred him to Buzz Bennett, who created Jesse's name. Jesse was married to a professional stripper and always encouraged the staff to go out and see her work.

Harry Scarborough (9-12pm) arrived from KYNO Fresno in June to replace Bob Foster who went to KIMN in Denver. Harry would soon move to the 6-9AM shift after Jim Carson left for KFRC. Harry was well suited for this slot, encouraging his listeners to 'rise and rejoice' to open the new day. He was known as a free spirit, often high strung, nervous and fidgety. Harry went over to KCBQ in 1971, and is now out of radio.

Peter Huntington May (9-12PM) arrived in August, probably from KIKX in Tucson where he was on staff in 1968, to take the shift Harry Scarborough had vacated. His career extended back to 1961 when he was at WDGY in Minneapolis.

KGB was nearing an end of an era. Buzz Bennett was totally dominant using his considerable skills to perfect the Drake format. His performance merited the PD position at KHJ when it opened up in August 1970, but RKO management dropped him from consideration when they saw his long hair, t-shirt, and spurs. Buzz's lifestyle included motorcycles and drug use which did not fit the straight edge Drake image. To make a point, Buzz resigned immediately from KGB, and for revenge, he accepted the program director position at rival KCBQ in December 1970 and started recruiting Drake staff. Always a favorite among the disc jockeys, many followed him to KCBQ.

Charlie Van Dyke was hired by General Manager Bill McDowell to replace Buzz Bennett as Program Director. The 23 year-old arrived from KFRC in August 1970. He had informed Drake that he was going to quit if he didn't get the KGB PD job. He brought a more mainstream image with a showbiz sparkle behind programming and promotions. He kept a condo with a view overlooking the city that served as party central for the staff. His admirers describe Charlie as a moral builder who kept things upbeat, positive, and fun every day. He was one of only seven to have worked for the three big Drake-Chenault Stations, and the only one to be PD of all three.

Barry Kaye (Barry Keller) (6-9PM) arrived in September from KTSA San Antonio after Rich Brother Robbin left for WDRQ Detroit. Bill Drake had told Charlie Van Dyke to check into thus guy in San Antonio who did this odd noise and was winning his time slot. He was the first DJ hired by Charlie Van Dyke and even participated in Charlie's wedding. Barry, by profession and preference, was a singer who performed under his real name. The name Barry Kaye was derived to separate his roles of singer and DJ. Barry used a signature "aaoogha" (like a Model T horn) to distinguish himself to listeners. Upon his arrival to San Diego, he was met at the airport by Harry Scarborough and Bobby Ocean, pretending to be Bill Drake and Bill Watson. He shook up management on his first night with 'stage fright' that kept him from speaking. Gene West came in and

KGB RADIO AM 1360

helped him through it. A bottle of whiskey became a fixture in a hidden drawer in the control room. During his tenure with KGB, he had a hit, 'On My Way' in November 1971 that charted nationally. Barry was heavily promoted throughout his KGB tenure. He was the featured DJ on a TV ad for the station and he appeared on a survey nearly every month he was at KGB. When Bobby Ocean became PD, Barry was offered the Noon-3 shift, but he disliked the direction the format was heading so he quit and returned to station KILT Houston where he had been previously. Despite his San Diego stint, his base has been in Houston, TX from the beginning of his career. He has been with oldies station KLDE in Houston since 1994.

The station continued to publish special editions of surveys. In September, a *Hall of Fame* survey was published of top hits from 1955 to 1970. Photos of the DJs were on the back. A Bobby Ocean/Vick drawing of the DJs clustered within Roman columns is on the front. The *Top 100 of 1970* was published at the end of the year that featured a collage of station staff and activities on the front and a group photo of the DJs on the back, similar to the *Hall of Fame* format.

1971

The station had devastating losses in talent that could not be replaced as the year began. With rival KCBQ hiring nearly all the KGB staff, by now, KGB was the only Drake station not winning its market. The station was giving away large cash prizes, boats, and even a Rancho Penasquitos tract home! All stations were required to maintain a Public File that contained the operation license, and legal claims, station violations, FCC visits, public complaint letters and so on. Buzz Bennett and Chuck Browning were coming over from rival KCBQ to view the station Public File at the front desk, while PD Van Dyke could only watch and fume. The lines were cut at the unmanned KGB transmitter site at 52nd and Kalmia on the first night of the ratings sweep to the gain of KCBQ. Buzz even found a way to inflate Arbitron ratings diaries. On the bright side, the KGB staff found a way to speed up the turntables using splicing tape to respond to the KCBQ format of speeding up songs. Here were the replacements.

Big John Carter (John Yount) (6-9AM) grew up in San Diego and attended Mission Bay High School. As a youngster, he was not alone in watching the KCBQ jocks in the studio window at 7th and Ash and deciding that was his future. He arrived in February from KYNO Fresno to replace Harry Scarborough who went to KCBQ. Big John was a graduate of the Bill Wade School of Broadcasting and had also worked at KACY in Oxnard under the name of Spanky Elliott. He left in October to return to Fresno at KFIG.

Shotgun Tom Kelly (Tom Irwin) (9-Midnight) began his radio career at KDEO as a seventh grader working the boards and being a gopher for the newsroom after school. His first on-air experience came in 1966 on weekends at KPRI in San Diego. He arrived in February

from KAFY in Bakersfield to replace Peter Huntington May who went to KCBQ. The name 'Shotgun' was applied by fellow DJ Dave Conley when Tom, using the name of Bobby McAllister, (which he hated), worked at KACY Oxnard. Dave Conley moved on to become PD at KAFY Bakersfield and invited Tom to join him. Tom wanted to use his real name but the station GM didn't like Irwin. They considered Collins, but it sounded like a drink. They considered Carson, but it sounded like a late night TV host. Then they considered Tom's Scot-Irish ancestry and Kelly was born. As a native San Diegan, Tom's goal was to jock at Boss Radio KGB. Tom had sent a tape to Charlie Van Dyke when he heard that Peter Huntington May was leaving. After listening to the tape, Van Dyke went to Bakersfield to listen to Tom on the air and offered him the job. His tenure at

KGB was his big break into the business. This would not be the only tour with KGB or KCBQ for Shotgun.

Johnny Mitchell (Paul Stelljes) (3-6 PM) arrived in March from KYNO Fresno where he had been PD and did the afternoon drive shift under the name of Harry Miller. Buzz Bennett had contacted him to come to KCBQ. But being loyal to the Drake organization, Johnny offered them first shot and they placed him in the afternoon drive at KGB. 'Johnny Mitchell' grew out of convenience when a spare jingle, prepared earlier for Johnny Mitchell at KHJ, (and also known as Sabastian Stone at KFRC and Johnny Holiday at KCBQ), became available from Johnny Mann. Johnny Mitchell served as "fill-in" Program Director while Interim PD Bobby Ocean "programmed" the station from a Fort Ord stockade via telephone. He served as Music Director when Ron Jacobs arrived, but hated the working conditions and found his way to KFRC with better management and equipment. He remains a close friend of Bobby Ocean to this day.

Mark Richards (Michael Spears) (Noon-3PM) arrived in May from WYSL, Buffalo, NY, where had been PD, after Gene West got the job he always wanted at KFRC San Francisco. Mark was hired by Charlie Van Dyke. The two had worked together at KLIF Dallas, TX, where Mark was known as Hal Martin. The name 'Mark Richards' grew out of the availability of two jingles that were spliced together that formed the name. But this association would only remain until June 19. Drake considered him a bad broadcaster and ordered Charlie to fire him while Mark was on his honeymoon. Charlie did so with class by meeting them at the airport with a limo and flowers for Kathy. Despite this setback, Mark has had a very successful broadcast career. After KGB he went to CKLW. He returned to Dallas in 1972 and launched the

city's first live FM station KNUS. He served as PD at KFRC San Francisco and twice won PD of the Year. His stations have won *Billboard Magazine's* Station of the Year three times. His website reminds visitors that "Life's short. Celebrate the joy and do for others as often as you can."

KGB RADIO
June 1971 - March 1972

The Drake radio network began experimenting with the format and decided to drop all references to "Boss", the name they had branded their format since the beginning.

The weekly survey became an un-numbered *KGB 30* to promote the station identity. Music selection and format remained the same. The station conducted aggressive contests that involved giving away a house, a boat, cars, and cash. Staff changes occurred regularly, highlighted by a frequent overturn in PDs and a Viet Nam era act of defiance.

Charlie Van Dyke (Chuck Steinle) (Noon-3PM) originally studied to become a Jesuit priest until he had the opportunity to fill in one day as a DJ at KLIF in Dallas, Texas. He was advancing through the Drake organization when he took over PD responsibilities in August 1970. He returned to the airwaves in June 1971 after firing Mark Richards. He held this shift until Gene West returned from KFRC in August. Charlie then took the 6-9 AM shift in October when Big John Carter left. As ratings sagged Charlie was fired and Bobby Ocean took over as Interim PD and filled his shift. Charlie went to 93 KHJ in 1972 and developed his career at various stations concluding in 1998. Cycling back to his roots, Charlie renewed his pursuits in the Catholic Clergy and was ordained a deacon to serve the Church 30 to 40 hours per week without pay. He is a voice-over artist and resides in Paradise Valley near Scottsdale, AZ.

KGB RADIO AM 1360

Chuck Browning 'The Chucker' (9-Midnight) arrived in July following a stint as the morning man at KCBQ, where he was fired for playing a Beatles song that he was told not to play. This shift opened when Shotgun Tom was told by Bill Watson that he did not have enough experience and was being transferred to KYNO Fresno. Shotgun declined and quit. Chuck stayed only a brief time before being fired when he gave away a cart machine on air after it failed to operate properly after repeated attempts. He then reunited with Buzz Bennett at KUPD in Phoenix. He was one of only seven who worked at the three big Drake-Chenault stations. His popularity among DJs was reflected in *A Tribute to the Chucker* in 1990 following his death on March 3, 1988 from lung cancer. Jesse Bullet followed 'The Chucker' in this shift.

K.O. Bayley (9-Noon) arrived in July to replace Christopher Cane who went to KCBQ. K.O. had been with KCBQ in 1970 before the Great Purge. This was his second tour with KGB. He was on staff in 1966-67 under the name of Bob Elliott and he was now considered an 'old timer' by his colleagues, who thought he was scary to some degree. K.O. enjoyed whiskey, cigars, and Pall Mall cigarettes. He was a boxer while in the Marines, and later competed professionally. His favorite past time was going bowling, getting drunk, then picking fights with Marines.

Gene West (Noon-3PM) returned by August to his old shift after he and KFRC PD Paul Drew could not get along. Gene had worked the overnight shift there.

Bobby Ocean (6-9AM) returned in October after quitting the business, (digging ditches did not pay the bills). He replaced Charlie Van Dyke as PD on an interim basis, but soon 'programmed' the station from a telephone booth at the Fort Ord US Army Base. He had been attending National Guard meetings with hair just long enough to 'touch the collar' contrary to military rules. The Captain fumed, "We'll just send you to Viet Nam, boy!" To which Ocean replied, "You'd have me killed because of my hair?"

"That's right, son" he replied, which further set Ocean's defiance. Instead of being sent to Nam, he was merely placed in the stockade for a few weeks, and would place his phone calls when not cleaning barracks, moving lockers, painting, etc. His interests were more aligned with *I Ching*, (the Chinese Book of Changes i.e. fortune telling).

Don Dale (12-6 AM) left in June and Jesse Bullett moved to this shift.

1972

As the year opened, Bobby Ocean was Interim PD, but was still a guest of the US Army in a stockade at Fort Ord. Johnny Mitchell was now serving as fill-in PD. In the meantime, the Browns were working on a plan to bring Ron Jacobs in as PD. The top 40 era of KGB was about to end as a "progressive" rock era was about to begin, but not before more staff changes occurred. The final line up was as follows:

Bobby Ocean (6-9AM) Bobby Ocean remained until leaving to KFRC in September. In addition to offering wit and innovation to the airwaves, he has been a respected industry leader and still performs voice overs through his firm in San Rafael. Ironically, he attends the same church that Gene West's Mother has attended. Ocean is still heard on KFRC-FM in San Francsico.

Paul Stelljes (9-Noon) (Paul Stellgis) returned to his real name after Ron Jacobs decided that 'Johnny Mitchell' sounded too much like a Top 40 air name.

Gene West (Noon-3) had just married and was planning a honeymoon cruise, when new PD Ron Jacobs informed him that he would be switching shifts with Tony Mann and be working overnight. Gene felt this was a demotion and quit. He moved over to KSEA briefly while Rich Brother Robbin was setting up K100 in Los Angeles. He joined Rich at K100, but gradually became disenchanted with the instability of radio employment. He left radio for electronics equipment sales. That eventually gave way to a return to school to complete two masters programs. He has been teaching for 15 years, and is currently an administrator, in the LA Unified School District. He is at a school in a low income, poor Latino area, where he is known only by his real name.

K.O. Bayley (3-6 PM) could not adapt to Ron Jacob's 'Recycled' radio format, which resulted in numerous arguments, and eventually his firing, despite the mutual respect between the two. He landed at KILT in Houston, (with Barry Kaye), by the end of the year. He eventually left radio and began painting houses.

Whenever a listener would call in and ask K.O. about whatever happened to a DJ who had left, he was known to tersely reply, "he got killed in a plane crash." Ironically, though he never learned to drive, K.O. was killed in a Michigan traffic accident in 1978 caused by a drunk driver.

Shotgun Tom Kelly (6-9 PM) Shotgun Tom Kelly participated in a walkout at KCBQ on January 4 that resulted in PD Johnny Mitchell inviting him back to KGB. He replaced Barry Kaye who quit in January, but Shotgun left again in June when new PD Ron Jacobs fired a number of staff. He joined Buzz Bennett at KRIZ Phoenix.

Jesse Bullet (9-12 PM)

Tony Mann (12-6 AM) arrived from KFXM in San Bernardino. He was best known from KRIZ in Phoenix. Those that recall him from KGB do not do so fondly.

Ray Cooper (Weekends)

The staff was either fired and collected a severance through June, or were invited to remain and work within the new format. But the environment was very different, the music was very different, and the expectations seemed impossible to meet. Only one 'Boss Jock' remained on staff past June.

RECYCLED ROCK
April 1972 - 1975

Ron Jacobs was hired as program director in February 1972 to move the station from dead last in the market to a ratings winner. As part of the package to bring Ron to KGB, the Browns agreed to purchase land in Hawaii that Ron wanted. Upon arrival, he surveyed the marketplace and moved the station toward a format similar to FM Underground by broadening the play list to a variety of genres and relying less on national trends and sales. The 'Boss Jocks' were on payroll until June to provide time to find other employment, but a different crew began in April that brought a different outlook on music and broadcasting. He brought in Rick Leibert as a full-time Assistant PD. Jacobs was credited as the program genius at KHJ during the Boss Radio years and actually made the Drake format work. The Browns had been absolutely sold on Jacob's programming abilities for years and gave him exclusive control over station operations. A strict task master insisting on button-down professionalism from his staff, his mantra was, "Be excellent and I won't fire you." He was a fearsome and feared employer who devoted all his time to making the station achieve his vision.

A change in focus, from air personalities to productions, began on April 12. The station introduced a new format that included a critique on the evolution of Top 40 radio formats, particularly Boss radio. Like Buzz Bennett, Jacobs was an admirer of Tofler's *Future Shock* and it became required reading of the staff. New emphasis was placed on rock histories, documentaries, quasi documentaries, and promotions. Gone was the weekly survey of the top singles. A 'pinup' of the top 15 albums emerged in the *Rock'n Thirteen-Sixty KGB* survey beginning on May 23. The Pinup was produced bi-weekly. New names emerged on the air: Cap'n Billy (10 AM-2 PM) Jean-Paul (2-6 PM), Ernie Gladden (6-10 PM), Gabriel Wisdom (2-6 AM), and Digby Welch. At first, the play list was virtually unlimited with over 6000 titles that tended to center on AOR tracks, classic rock, and breaking artists, (such as David Bowie at the time). KGB AM, and its long-time sister station KGB-FM, began a simulcast on August 1,1972. This presented listeners who only had AM receivers to experience the sounds of progressive FM rock. A charity ball was held at San Diego Stadium on November 12, 1972. It attracted more than 51,000 fans to benefit the United Way.

The air staff during 1972 included the following:

Cap'n Billy (Bill Hergonson) (10 AM – 2 PM) graduated in TV/Radio from Ithaca College and had worked at Watermark with Ron Jacobs. He arrived in May 1972 from KRLA and is responsible for the advent of the Home Grown album series of local artists. He moved to the 6-9 AM shift by August as the FM simulcast began. He moved on to KAFM in 1974.

Jean-Paul (2-6 PM) (Paul Stelljes) set a new record with a third air name at the same station. He "escaped" (as he put it) to KFRC in June using the name 'Eric Chase'. Paul has returned to California from Houston, TX in June 2003, and has returned to his air name of Eric Chase at KRTH in Los Angeles.

Ernie Gladden (6-10 PM)

Jesse Bullet (10 PM-2 AM) was not a likely holdover 'Boss Jock'. But he made his case to Ron Jacobs. The job had been intended for Bill Hergonson who was doing much of the production work and had been promised a shift.

Jesse became PD of the automated KBKB-FM in an oldies format in 1975. In 1976, he moved to KDEO when it switched to an album format. He consulted for radio stations for a number of years using his real name of Richard Bullen. He was general manager of KEZN in Palm Desert, California before retiring to Lake Tahoe.

Gabriel Wisdom (Ben Wool) (2-6 AM) was well known to San Diego listeners before his arrival. His career began in 1968 at 'Free Form' KPRI-FM and remained with 'progressive rock' oriented station KGB and KMET in Los Angeles. He arrived at KGB in 1972. As DJs were partying in Bullet's home, Bullet got a phone call and announced that, "it was Gabriel Wisdom, the Silver Surfer and 'coolest far-out DJ in town," was coming to KGB. It was The Bullet who had introduced Wisdom to Ron Jacobs. He now lives in Rancho Santa Fe.

Wisdom now a syndicated investment advisor on Business Talk Radio. He is also Managing Director of American Money Management LLC that manages retirement plans.

PD Ron Jacobs wanted to put some pizzazz into the 9-10 AM hour of the day, so Cap' n Billy started adding songs with a local twist as shtick to fill the hour. The station wanted to do another benefit concert, but when the fire department imposed seating restrictions, it made another rock concert impractical. By now Cap'n, Billy had amassed a collection of tapes by local artists and the station converted the best of these into another promotion to benefit the United Way. Christened the *Home Grown* series based on Linda Ronstadts' first solo after leaving Stone Poneys, titled *Hand Sown, Home Grown, Perfect*. These albums were originally priced at $1.01 and were sold by the tens of thousands. Cuts from Home Grown albums were limited to promotional use only on the air.

Rick Leibert became PD in 1973 when Jacobs returned to Hawaii. Rick won the prestigious PD of the Year Award in 1975. He followed the Ron Jacobs tradition of maintaining iron clad control over station operations. Leibert conceived of the KGB Skyshows at San Diego stadium that continue through 2003. Leibert started his own marketing company in the late 1970s.

Former KDEO DJ 'Sunny' Jim Price arrived as General Manager in 1973 to replace Bill McDowell. This occurred as a result of a defamation suit brought against KGB by a local ad agency that KGB lost. McDowell was assigned the blame and fired.

Meanwhile, the format was experiencing problems. Without the focus of a tight play list, the format became unruly so the AM and FM broadcasts were split in 1974. KGB-AM returned to Top 40 as KGB-FM moved toward progressive rock.

Price recalls that, "We did a billboard campaign to inform the people of the difference. There were two eggs-one in psychedelic colors to represent the FM, and another in more neutral tones to represent the AM.-and in the middle there was a chicken atop the phrase, 'Non-identical twins.'[2]

"I wanted to make the most of this promotion, so I had a meeting with (program director) Ron Jacobs and his assistant, Rick Leibert, and we eventually came up with the idea of dressing up a guy in a chicken suit."

A journalism student was recruited from San Diego State University named Ted Giannoulas. Following appearances at the San Diego Zoo, Wild Animal Park, and baseball games at San Diego Stadium the concept took off and the KGB Chicken was born.

The following staff were on the air by 1974. The list of DJs is also not complete:

Chuck Clemens (6-10 AM) returned to the market in 1974 from KMEN in San Bernardino. He had been a stock broker for six years before that. He developed a Tyrone the Frog character for his morning show. He retired from radio after his time at KGB.

The station adjusted the format to a very mellow pop sound in 1975. Airchecks of that period reveal DJs frequently announcing the calls, (1360 instead of 136) but no jingles. Air time was weighted toward ads instead of music.

As the nation celebrated the bicentenniel in 1976, the station inaugurated the KGB Skyshow, originally at Mission Beach, but in later years at the San Diego Stadium.

Jim McInnes arrived in 1974 from KPRI where he used the name O.B. Fillmore. He held the longest tenure with the station, mostly on the FM side and well beyond the AM days, until he was fired by Clear Channel in 2002.

Wizard Lew Rogers (Lew Ruggio) has been a fixture at KPNX Channel 12 in Pheonix since 1985.

Kevin McKeown began his career as a student DJ at Yale and arrived from WPLR in New Haven, CT. He left for KROQ in Los Angeles in 1976 where he eventually became general manager. He moved on to radio management and advertising before settling in as recording studio owner in Santa Monica. He became a community activist on homeless issues and a consultant to the local school district. He participated in numerous civic committees and was elected to the Santa Monica City Council where he now serves as Mayor Pro-tem.

Arthur 'The Vulgar' Boatman arrived from KPRI and left a lasting impression by showing up three minutes late to a staff meeting and getting fired on the spot by PD Ron Jacobs.

Larry 'The Cruiser' Himmell typified the mellow FM DJ who explained the album and the cut as he played each song. He arrived from KDEO in 1977 after they did a similar AOR format.

Bill Minckler arrived from KIMN in Denver and left for KYAS-FM. He is now the Director of Programming for Clear Channel radio network.

13K ERA
1979 - 1982

GM Jim Price fired Liebert as program director and brought in John Lander. Lander felt that the station name needed to be changed to bolster the

KGB RADIO AM 1360

AM identity that had been taken over by KGB-FM. He adopted a traditional popular music format on  October 8, 1979 with a 13K service mark. The station was identified as 13KGB on the hour to meet FCC requirements. With a popular music list, greater emphasis on production value, aggressive (and over budget) contests, and imaging that was modified seasonally and for holidays, 13K rose to the top of the market for both AM and FM stations with a 6.4 share by their third book.

Local schools were blanketed with 13 inch rulers announcing "13K Rules." The station reintroduced the weekly survey in October as the 13K Connection of singles and albums using the same tri-fold design as the old Boss 30. The cover featured a recording artist instead of a DJ, and contained the lyrics of a current hit. Surveys often had a DJ photo. Keith Danon was hired as a roving DJ to play records at special events in the area while promoting 13K. The station returned to being a ratings leader with a Top 40 format as rival KCBQ began experimenting with a Country and Western format. The emergence of the Mighty 690 XETRA (formerly XEAK) with a CHR format led to the permanent departure of KGB from the TOP 40 format in 1982. The air staff had a strong tie to KYA San Francisco that included:

1979

Larry 'Cruiser' Himmell (5:30-9:00 AM) remained on staff a brief time.

John Lander (5:30-9:00 AM) did the Lander in the Morning show while serving as PD.

Chris Cane (9:00 AM–1:00 PM) The same Christopher Cane that was on the KGB staff from 1969 to 1971, then at KCBQ from 1971 through 1973, arrived from KMJC. He had been at KYA during the mid 1970s.

Dean Goss (1:00-5:00 PM) arrived from KCBQ.

Gary Cocker (5:00-9:00 PM) arrived from KTNQ in Los Angeles where he worked with KCBQ alumni Dave Conley.

Gary Knight (9:00PM-1:00 AM)

Barry Ryan (1:00-5:30 AM)

J.B. Mitchell was on staff for a brief time.

Chris Lance (Weekends) from KYA or KMJC.

Jeff Prescott spent his entire career in the San Diego radio market. He joined KGB-FM in August 1975 as a news reporter then as new director. He was news director at 13K.

1980

Brown Broadcasting moved into a new state-of-the-art facility studio in March 1980, designed by engineer John Barcroft, and located at 7150 Engineer Rd.

PD John Lander took himself off the air for the year in March after teaming with Jeff Prescott in the morning show titled "*The Breakfast Flakes*".

The line-up was adjusted as follows:

Gary Knight (5:30-9:00 AM)
Chris Cane (9:00 AM – 1:00 PM)
Dean Goss (1:00-5:00 PM)
Gary Cocker (5:00-9:00 PM)
Rick Gillette (9:00 PM-1:00 AM) arrived from KAFY in Bakersfield and was the replacement for John Lander's shift.

Barry Ryan (1:00-5:30 AM)

Jeff Prescott continued as news director, while **Chris Lance** and **Jon Driscoll** (left) handled weekend shifts.

1981

John Lander resumed the morning show with Jeff Prescott. But the yellow rose of Texas was beckoning Lander and he left in May. Rick Gillette was also not on the air at this time. Lander He can now be heard on WBMX *Mix 94.5* in Boston. The beginning year line-up was:

John Lander and Jeff Prescott (5:30-9:00 AM) Jeff expanded his news activity to team with Dean Goss on the morning drive show. To this day, Jeff continues to handle news for KOGO during the morning drive.

Chris Cane (9:00AM–1:00PM)

Dean Goss (1:00-5:00 PM) Dean went on to host or co-host television game shows before returning to radio. He now does the morning drive at KFRC in San Francisco.

Gary Cocker (5:00-9:00 PM) moved to this shift after Rick Gillett left.

Gary Knight (9:00 PM-1:00 AM)

Barry Ryan (1:00-5:30 AM)

Tommy Lee (Tommy Sarmiento) (Weekends) was hired by John Lander to serve as Production Director, but eventually ended up doing weekends on the air. He stayed with KGB-FM as Production Director from the format shift to all-news in 1982 until 1992, when KPOP was inaugurated. Preferring the pro-

duction side to radio, Tommy won the Humboldt Award in 1992 for his work on a Playco commercial. Tommy now applies his computer talents to training high school students for their A+ certification in computer technology at a San Diego area high school. He enjoys marathon cycling during any free time he can find.

Casey Kasem (8:00-Noon Sundays) American Top 40 **syndicated** national broadcast

After Lander left in June, Gary Knight was moved to the morning drive and Rick Gillette returned to take his shift. But within a month, Knight returned to his night shift and the other shifts were adjusted as follows:

Dean Goss (5:30-9:00 AM)

Chris Cane (9:00AM–1:00PM)

Gary Cocker (1:00-5:00 PM)

Rick Gillette (5:00-9:00 PM)

Gary Knight (9:00 PM-1:00 AM) left by September.

Barry Ryan (1:00-5:30 AM)

David Good (Weekends)

The frequency of shift changes due to staff departures underscores the significance of John Lander. The shifts changed again in September when Gary Knight left:

Dean Goss (5:30-9:00 AM)

Chris Cane (9:00AM–1:00 PM)

Gary Cocker (1:00-5:00 PM) Gary moved to the FM side of KCBQ under Bob McKay as that station went to a country format. He remained there until 1984 when he went to KFMB to stay another six years. Gary now performs production work for XETV in San Diego.

KGB RADIO AM 1360

Rick Gillette (5:00-9:00 PM) became the program director when John Lander left.

David Good (9:00 PM-1:00 AM) replaced Gary Knight.

Barry Ryan (1:00-5:30 AM)

The shifts in November after Gary Cocker left were:

Dean Goss and *Jeff Prescott* (5:30-9:00 AM)

Chris Cane (9:00 AM–1:00 PM) would leave by the end of the year.

David Good (1:00-5:00 PM)

Rick Gillette (5:00-9:00 PM)

Bob West (9:00 PM-1:00 AM)

Barry Ryan (1:00-5:30 AM)

1982

The final shifts consisted of the following:

Dean Goss and *Jeff Prescott* (5:30-9:00 AM) Dean went to KRTH when the station went all-news. Jeff remained as news director of KGB-FM until 1990.

Ted Wayne (Ted Ziegenbusch) (9:00AM-1:00 PM) Ted attended the Don Martin Radio School in Hollywood that led to his first radio job at KMEN in San Bernardino. He came to the San Diego market in 1980 from KLAV Las Vegas to join the staff of 'The Mighty 690' XTRA. He moved to 91X that same year before coming to 13K to replace Chris Cane. Ted now does the morning drive at Christian Contemporary station KFSH-FM in Anaheim, CA.

David Good (1:00-5:00 PM)

Rick Gillette (5:00-9:00 PM) had the distinction of being the last DJ on the air.

Bob West (9:00 PM-1:00 AM)

Barry Ryan (1:00-5:30 AM)

The weekend staff were *Ashley Gardner, Chris Lance,* and *Cliff Roberts.*

The last surveys were published in February as a major format change was about to occur. *When the Music Stops, It's News* aired during the closing days of the 13K format that chronicled the history of KGB from 1959. The program included interviews with former staff, airchecks, and commentary describing all the formats of the rock era.

KCNN 1982

This was an attempt to air a live feed of CNN Headline News beginning on March 15, 1982. But the news anchor would refer to images on the screen. The concept was flawed from the start. Don Howard of KCBQ fame had a weekend big band nostalgia show that was highly popular. Eventually, the KCNN format was dropped and the Big Band concept became the main staple of the format. The calls of KPOP were sought but were unavailable at the time, so KPQP was licensed by the FCC about 1984. Eventually, the KPOP calls became available.

KPOP

1360 kilocycles on the dial will no longer produce a Top 40 sound. Instead, nostalgic sounds of the big band area, popular, and jazz music will be heard along with DJs who were on the scene during the Boss Radio era. Most notably among these is Happy Hare (Harry Martin) who was a titan of the airwaves of KCBQ in the 1950s, 60s, an early 70s.

FOOTNOTES

[1] *The Hits Just Keep On Coming*, Ben Fong Torres, 1998, p 91

[2] San Diego Magazine, January 1984, p 151

Images – KGB Radio AM 1360

KGB Studio and Transmitter – 1929.

KGB Studio at 4146 Pacific Coast Highway – 1955-1979.

Former KGB Studio in 2003.

KGB Transmitter at 52nd & Kalmia

KGB RADIO AM 1360

The Weekly Surveys

These 1963 surveys evolve from only a playlist in Spring to promotions, personalities, and the playlist by Fall. The station was an ABC affiliate with no footing in the Top 40 market.

Early surveys of the Drake format as Boss Radio develops. The Beach Boys concept, unique to San Diego, was used in 1964 and 1965.

Debut – May 1964

Debut – August 1965

Debut – February 1966

KGB RADIO AM 1360

Debut – October 1966

*Debut – April 1968
(DJ Drawing)*

*Debut – November 1969
(DJ Photo)*

Debut – June 1971

OFFICIAL KGB 30

PLAY LIST — PREVIEWED 3/6/72

1. PUPPY LOVE — Donny Osmond
2. HORSE WITH NO NAME — America
3. MOTHER AND CHILD REUNION — Paul Simon
3b. ME AND JULIO — " "
4. NO ONE TO DEPEND ON — Santana
5. HEART OF GOLD — Neil Young
6. JOY — Apollo 100
7. JUNGLE FEVER — Cha-ka-chas
8. THE LION SLEEPS TONIGHT — Robert John
9. HURTING EACH OTHER — Carpenters
10. WITHOUT YOU — Nilsson
11. BLACK DOG — Led Zeppelin
12. ROUNDABOUT — Yes
13. PRECIOUS AND FEW — Climax
14. WE'VE GOT TO GET IT ON AGAIN — Addrisi Bros.
15. BANG A GONG — T. Rex
16. DOWN BY THE LAZY RIVER — Osmonds
17. CRAZY MAMA — J.J. Cale
18. WITCH QUEEN OF NEW ORLEANS — Redbone
19. EVERYTHING I OWN — Bread
30. JUMP INTO THE FIRE — Nilsson
31. THE FAMILY OF MAN — 3 Dog Night
32. SIMPLE SONG OF FREEDOM — Buckwheat
33. ROCK AND ROLL — Led Zeppelin
34. CASTLES IN THE AIR — Don McLean
35. THE FIRST TIME EVER I SAW YOUR FACE — Roberta Flack

Debut – March 1972

Debut – May 1972

KGB RADIO AM 1360

Debut – November 1979

KDEO Raydeo AM 910

Through several ownership changes, KDEO offered an entertainment-driven approach to broadcasting. Programming relied on the strength of individual personalities to attract and hold an audience. This individualistic approach left the station without a definitive sound or identity, but did enable air staff to freely develop and display their talents. One former KDEO PD described the setting as a 'Parade of Clowns'. No one considered KDEO to be the pinnacle of their career, but many considered it as their launching pad.

1959
A Dandy Year of Beginnings

Dandy Broadcasting acquired the station and hired Sam Babcock as Program Director to inaugurate a Top 40 format. Babcock hired a staff and adopted the 'Color Radio' theme that was being used successfully at KFWB in Los Angeles. Gimmickry was no stranger to KDEO. The morning man was named 'Coffee' and the night man was named 'Shadoe'. Overnights were handled by a female DJ. The line-up included the following key staff:

Coffee Jim Dandy (Jim Washburne) (6-10 AM) was a bearded 'hip' DJ who arrived from Washington DC in the summer of 1959. His father was famed band leader Country Washburne. Jim Dandy followed Buck Herring and Ted Millen in this shift beginning in January 1959. Millen moved on to the ABC TV affiliate in San Diego.

Sam Riddle (10 AM-2PM) arrived from Pheonix to replace Sam Babcock when Sam became the full time PD. He was bubbly, chatty, and possessed a squeaky sounding voice owed to his youth. He was on staff a short time before moving to KRLA Los Angeles, then later to become more  famous at KHJ Los Angeles and the host of Boss City on KHJ-TV.

Mel Hall (2-6 PM) began his radio career while serving in the Army at Camp Rucker in Alabama. The soldiers built and operated station WVIK. Harry 'Happy Hare' Martin had preceded him there. Mel took a job at WABB in Mobile, AL after his discharge. After two years at KOCS in Ontario, CA, then  freelance radio and TV commercial work, Mel moved to KLAC in Los Angeles during their 'FutureFonicSound' experiment that resulted in everybody being fired. He was hired in December 1958 by Sam Babcock and served as Production Director and did the afternoon drive shift at a salary of $135.00 per week. Mel characterizes his early style as sarcastic.

Shadoe Jackson (Jerry Swearingen) (6-10 PM) arrived from Lincoln, NB and was appointed music director. Dandy Broadcasting bestowed the name 'Shadoe' upon him after hearing of another 'Shadoe' at a Midwest radio station. He was not glib, and lacked clarity in his delivery at the time. But the teen audience reacted well to this night time 'shadoe'. This caught the attention of local giant KCBQ, where he went in May 1960.

Sie Holliday (Shirley Schneider) (11PM-6AM) was the first female Top 40 DJ in San Diego and later at KRLA in Los Angeles. Sie was hired about June 1959 when the station began doing 24 hour broadcasts. She  had been on the air in Texas, but this was her first Top 40 station. She highly suspects that the deciding factor in her hiring was that they could pay her less than a man. She was fired when Tullis and

KDEO RAYDEO AM 910

Hearne bought the station during the summer of 1960. She now resides in Wichita Falls, Texas where she is an actress in live theatre.

1960
Tullis and Hearne Acquire Station

Mirror mirror on the wall . . . KDEO is the fairest of them all . . . (jingle).

Mel Hall became Program Director in early 1960 shortly after Sam Babcock left for KDAY in Los Angeles as PD. Sam died soon thereafter from complications due to an injury while in the Army. DJs came and went during the year. Tullis and Hearne took over the ownership of the station during the summer when Dandy lost its financial backer and had to liquidate its assets. The Tullis and Hearne network stations issued the *Fabulous 40* weekly surveys measuring 10" X 4". The survey featured the top 40 songs plus 15 potential hits and the top 10 albums. The play list was based strictly on surveys conducted with local record outlets and shunned big market sales trends. Accordingly, the play list would include pop hits, middle-of-the-road, and country artists on a regular basis. DJs were listed below the record lists as The Musical Mob at KAYDEO. The bottom of the survey was used to promote DJs or a contest. Some of the greatest names in Top 40 radio that were hired that included:

Don MacKinnon (6-10 AM) arrived from KABC in Los Angeles in June 1960 to replace Coffee Jim Dandy when he moved on as program director at KRLA in Los Angeles. Jim was killed in an auto accident in 1965.

Don is considered by many as the most entertaining, fun, and inventive Top 40 DJ to ever grace the airwaves. He brought attitude, experience, creativity, and intellect that greatly influenced the other staff. The guys regularly walked to a nearby coffee shop for meals and banter. Don was known to grab his ankles and hop to the diner clucking like a chicken for laughs. He went on to acclaim at KEWB in Oakland. But he had a dark side as well. A heavy drinker, this contributed to his death in an auto accident in Malibu in 1965.

Dick Williams (10AM-2PM) was on staff by March 1960 and did the mid day shift. He left later in the year.

Don Bowman (10-2 PM) followed a childhood ambition to become a disc jockey. But his friendship with Waylon Jennings led him into song writing and comedy on the country circuits. He arrived early in 1960 and was designated as Program Director for about three weeks when Tullis and Hearne bought the station. Don was funny and talented, but at the time, he lacked an understanding about organization, and discipline behind the production aspects, that impeded his success as a PD. His show included a fictional character voice named Gruesome Goodbody. However, the local Goodbody Mortuary took great exception to Don's character. He soon left for KEWB in Oakland, CA. In February 1961, he had a hit on the charts titled *Coward of the Alamo*.

Mel Hall (2-6 PM)

Mike Ambrose (6-Midnight) arrived about March. He brought a deep resonant voice from legendary radio station KXOL in Fort Worth, TX. He moved about in shifts over his lengthy career at KDEO. As a DJ, his early style was oriented more toward announcer than entertainer.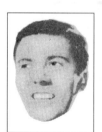

Dale Ware (Midnight-6 AM)

Chip Allen (weekends and part-time) arrived in early 1960. He quickly earned the nickname 'Comander' because his primary occupation was being in the Navy.

Dave Banks arrived from KQEO in Albuquerque, NM early in 1960 and stayed briefly. He was being shifted around by Dandy over personal problems and signed off with "This is Dave Banks saying Thanks".

Adam Story (Irv Weinstein) was news director, but his wife didn't like the area so they returned to their native area of Buffalo, NY. Raymond E. Spencer of KFXM, (and subsequently at KCBQ), followed him and inaugurated the present tense news delivery that the station was known for during their Top 40 run. Other news staff included Johnny Huddleston from KSON, (renamed J. Paul Huddleston by Mel Hall), and Sig Smith who left radio to become press secretary for Congressman Lionel Van Deerlin in 1962.

1961
Radio 91-derful Leads Market

Mel Hall left the station on March 10 to become PD at WJJD in Chicago. This opportunity emerged due to a recommendation made by Don MacKinnon. He returned to San Diego in early 1963 to work at XETV-TV. Mel wrote and produced the *'You belong in the zoo, the San Diego Zoo'* song. *He* returned to radio as PD at KRLA in 1965. Mel formed Cinira Corporation in 1966 to produce commercials and original music that enjoyed great success.

The station took on a Tulsa, OK bent under General Manager John Pace, Group Program Director Roy Cordell, station Program Director Noel Confer, and morning man Rex Jones. They had all worked at station KAKC in Tulsa. KDEO enjoyed their only ratings lead over KCBQ during portions of the 1961 rating period. This 1000 watt station prevailed over 50,000 watt KCBQ because of the signal coverage over downtown San Diego where the Hooper ratings were taken.

Noel Confer

There were several other reasons behind KDEOs success in 1961.

- KDEO was using a play list derived entirely from phone surveys of record outlets rather than industry trends. This resulted in a regular, and sometimes exclusive, mix of pop, folk, and country artists.
- The news was entertaining. As news director as well as program director, Noel Confer leaned toward lurid and scandalous stories and continued the practice of airing fast breaking present tense news by what would appear to sound like fearless reporters on the scene as events unfolded. This would aid in holding the music audience over the news breaks. It helped that Confer had live theater experience to help dramatize the effect.
- Competition was reduced. Although XEAK had been losing market share, their format change from Top 40 to all news in April resulted in only two local Top 40 stations.

Ken Dowe (6-10 AM) planned to enter the air force and train as a fighter pilot. On the day he was to enter, he followed a hunch and signed on as a disc jockey at a Mobile, AL radio station. From there, he came to KDEO.

Dick Williams (10AM-2 PM) went to CFPL Ontario, Canada by June.

Mel Hall (2-6 PM)

Steve Crosno (6-11 PM) started his career as KGRT in Las Cruces, NM at age 16. From there he went to KELP in El Paso, TX before arriving at KDEO. Although San Diego represented a substantial salary increase for him, Steve wasn't happy in the city by the bay. "I missed the people of El Paso and the climate." [1] So Steve returned to KELP by June. He has remained in El Paso ever since, deftly maneuvering through Top 40, Rhythm and Blues, and Tejano formats. The Texas House of Representatives passed a resolution on May 28, 2001 recognizing Steve for his contributions "to the culture and lives of countless people" as a radio personality.

Mike Ambrose (6-11 PM) replaced Steve Crosno in this shift.

Chip Allen (Weekends) survived the turmoil that comes with new ownership by being willing to work on any shift, at any time, for minimum wage. His earnings were said to accurately represent his radio talent. But it was a sure means to learn the trade.

KDEO RAYDEO AM 910

The shifts were adjusted after the arrival of new Program Director Noel Confer about June:

Ken Dowe (6-10 AM)

Mike Ambrose (10AM -2 PM) moved to mid-days.

Ed Thomas (2-6 PM)

Noel Confer (6-11 PM) arrived from XEAK about June when it dropped its Top 40 format for all news. He replaced Mike Ambrose in the evening shift when Mike went to mid days. Noel also served as program director.

'Wild' Bill Wade (Bill Wade) (6-11 PM) arrived from KACY in Oxnard in late 1961. Bill attended a radio school in Hollywood and got his first job by responding to an ad in the LA Times for an announcer at a station in Ridgecrest. Howard Tullis hired him to do the night shift.

'Smilin' Lee Shoblom (11-6 AM) began his career as a civilian electronics technician at North Island while studying to pass the FCC commercial licensing test. He was hired at KSON in 1957 as a night board operator. He worked his way up to become a DJ there before moving to KPRI-FM in 1958. He joined KDEO in 1959 and set his sights at trying to draw better ratings than KCBQ.

Chip Allen filled in on weekends and part time.

1962
Radio 91-derful Line-Up

The staff as of January were as follows:

Ken Dowe (6-10 AM) accepted an offer from Gordon McLendon for a job at KLIF Dallas by September. He has flourished since. He is now Chief Operating Officer of Service Broadcasting Corp. that owns K104, the number one station in Dallas.

Mike Ambrose (10 AM – 2 PM)

Ed Thomas (2-6 PM) filled Bowman's shift early in the year. He either left the station for the remainder of the year, or followed Lee Shoblom as 'Ben Bright' in the overnight shift.

Noel Confer (2-6 PM)

Bill Wade (6-11 PM)

Ben Bright (11 PM-6 AM) was the name of a fictional character applied to whom ever worked the overnight shift during the tenure of Noel Confer. 'Ben Bright the All Night Satellite' was actually **Lee Shoblom** at this time.

Chip Allen (Weekends)

The following changes occurred beginning in June:

Rex Jones (6-10 AM) had been program director at KAKC in Tulsa where he had worked with the present management staff of KDEO. He arrived by July when Ken Dowe went to KLIF Dallas. He joined Dowe at KLIF in January 1963. Rex died on June 29, 1988 in Dallas.

Mike Ambrose (10 AM – 2 PM)

Don Bowman (2-6 PM) returned about mid year from KEWB. He later went to KBBQ in LA, then became a comic on the Country Western circuit. He received the CMA Comedian of the Year Award in 1967,68, and 69. During this time, he appeared in several films that included *Hillbillies in a Haunted House* and *Hillbillies in Las Vegas*. He continued to write songs performed by Waylon Jennings and Willie Nelson. He was the host of American Country Countdown from 1973-78. He now lives in Branson, MO.

Noel Confer (6-11 PM) would become Two Gun Noel Kelly at Country Radio KSON. Noel now resides in Tulsa, OK. Ben Bright (11 PM-6 AM) was probably portrayed by 'Wild' Bill Wade during this time as he was about to move on to KHJ Los Angeles in 1963. This move preceded the Drake

'Boss Radio' era that would dominate the balance of Bill's on-air career.

Lee Shoblom (Weekends) replaced Chip Allen by July.

1963
Radio 91-derful Line-Up

Chuck Daugherty (6-10 AM) Chuck has enjoyed an extensive career that dates before Top 40 formats. While working at ABC affiliate WXYZ in Detroit, he did radio and TV stints on the original live *Lone Ranger* and *Sergeant Preston of the Yukon* shows. He also did regular network feeds of live band music, sports, and news.

When Top 40 emerged, as a DJ Chuck was instrumental in mainstreaming black artists and labels among white audiences. He also originated the record hop by being the first DJ to acquire a license to play dance music. Chuck came to California to fill-in for a DJ at KDAY in Los Angeles in 1962. He then moved to Tullis and Hearne station KFXM in San Bernardino. To drum up ratings, Chuck set out to break the marathon drumming record of 100 hours. After five days of continuously drumming, playing the drums while being driven in a VW bus throughout the region to promote the event, he broke the record by exceeding 101 hours. His record was broken in Britain two months later.

Chuck was transferred to KDEO from KFXM in January, replacing Rex Jones, who went to KLIF in Dallas. As an early promotion, Chuck began a contest to swap items beginning with a certified, hand-polished, paperclip. He swapped the paperclip with a kid for a homemade carpeted skateboard. DJs tried out the skateboard in the KDEO parking lot and declared it to be the 'Official KDEO Skateboard.' With that provenance came demand. It was soon swapped for a 21-inch RCA TV set, which was swapped for a ten-foot surfboard. This spiked up the ante considerably in surfing-crazed San Diego. Over time, the trading continued, landing a Cadillac Hearse, (named the KDEO Cad by staff), a BSA motorcycle, a mint 1939 Ford Woody station wagon, a 1938 Buick Opera Coupe, and finally a Class E Hydroplane that clocked 163 mph in San Diego Bay. Title was held for each item as it was acquired. But alas, KDEO did not want to look like brazen profiteers, so one more swap was made to complete the promotion. The hydroplane was swapped to the kid with the skateboard, in exchange for the original paperclip.

A memorable incident occurred one Sunday when Chuck mentioned on the air, "grunions are running and I'm going to be there." He was referring to a grunion hunt at Belmont Park in Mission Bay. (For non-San Diegans, a grunion is a small fish that seasonally swims ashore to spawn). An estimated crowd of 10,000 listeners showed up. This prompted local police to declare a "major emergency" and to dispatch some 50 police cars with officers in riot gear. Un-amused local officials passed resolutions censuring the station and KDEO reacted by firing Chuck the next day. He was rehired a few days later. Shifts changed frequently during this time. Chuck switched shifts with Don Bowman briefly in March, then both returned to their original times.

After leaving California, Chuck moved on to some east coast stations, then back to Detroit. He eventually owned several stations in Michigan, but had since sold them and now enjoys retirement.

Mike Ambrose (10 AM –2PM)

Don Bowman (2-6 PM)

Noel Confer (6-11 PM) was completing his last year before moving to Country station KSON where he would become Two Gun Noel Kelly. Noel now resides in Tulsa, OK.

Tom Clay (6-11 PM) began his career at several stations in the Detroit market. He arrived about June from KDAY in Los Angeles to replace Noel Confer who took himself off the air to focus on PD

KDEO RAYDEO AM 910

duties. Tom was noted for being a storyteller about daily life experiences. He played records between his stories. He is most famous for his 1971 top 10 hit *What the World Needs Now is Love* which he recorded when he worked at KGBS. The record was a re-release of the Jackie DeShanon hit that began by asking children to define bigotry and prejudice with tracks of back-up vocals and famous speeches of the 60s. "I love radio so much . . . I would pray that all-night man wouldn't show up so I could work another six hours." Tom moved on to KBLA in Los Angeles. He died of cancer in November 1995.

Ben Bright (11 PM – 6 AM) was probably portrayed by **Lee Shoblom** after Bill Wade left.

Ed Thomas (Weekends) returned in January to replace Lee Shoblom.

1964
Radio 91-derful Line-Up

Tullis and Hearne stations KDEO, KAFY, and KFXM developed a new survey format introducing a tiger mascot. The top 40 songs, potential hits, and top albums remained as features, but advertising was added to the back panel to underwrite productions costs. The new bi-fold survey measured 8 1/4 X 3 3/4 featuring all the DJs on the cover.

The station hosted a concert in July that included a number of groups at Westgate Park, then home of the San Diego Padres minor league baseball team. Somehow, the crowd was said to be induced to throw rocks and create a riot situation. Police had to be called to subdue the crowd and end the performance. One might guess these "shaggy singers" were the *Rolling Stones* or *The Who*. But not in San Diego. The group was English folk singers *Peter and Gordon*.

The air staff at the beginning of the year consisted of the following:

Chuck Daugherty (6-10 AM)

Mike Ambrose (10 AM-2 PM) was completing his run. He went to legendary radio station KFWB in LA to work part time until a prime shift opened up. Mike remained a San Diegan while working at Los Angeles stations KRLA and KFWB. He was also at KOGO radio in San Diego in 1970. He was popular weatherman Captain Mike at TV station KGTV in San Diego until his retirement in April 2001.

Mad Lad **Morton Downey Jr.** (Sean Morton Downey) (2-6 PM) arrived from KAFY Bakersfield and would soon move on to WFUN Miami enroute to becoming a TV talk show host. KDEO listeners could join his Mad Lad Marauders and receive a commission as a Colonel with a secret code. He went on to fame and controversy in syndicated television. Photo was more recent. Morton died on March 12, 2001 of cancer.

Robin Scott (6-11 PM) was a good natured guy that overcame the constraints of polio to have a long and successful broadcast career. He developed the character 'Gus Whatt' and would ultimately end up on Country and Western stations as Bob Jackson.

Mike Hunter (11 PM–6 AM) arrived by April to retire the Ben Bright character when Lee Shoblom left.

The line-up changed considerably by August when the station performed live broadcasts from Belmont Park and hired a bikini-clad young lady to hand out surveys.

Robin Scott (5:30-9 AM) took this shift when Chuck Daugherty moved to mid-mornings.

Chuck Daugherty (9-Noon) took this shift when Mike Ambrose left.

Buzz Baxter (Noon-3 PM) arrived from KFXM San Bernardino for this new shift.

Hal Pickens (3-7 PM) replaced Morton Downey. Hal switched shifts with Robin Scott by the end of the year before going leaving to Los Angeles stations KFWB then KBLA.

Larry Boyer (7-11 PM) arrived to take this shift when Robin Scott moved to mornings.

(Carlton) *Corky Mayberry* (11 PM-5:30 AM) replaced Mike Hunter. After his brief time at KDEO, he worked at numerous stations in the Los Angeles market, including KBBQ with Robin Scott (Bob Jackson) and now lives in Amarillo, TX.

1965
Tiger Radio

The line-up was adjusted by March after Hal Pickens and Corky Mayberry left. It featured the following staff:

Chuck Daugherty (5:30-9 AM) moved on to WNEW New York.

'Sunny' Jim Price (9-Noon) began his career at a small radio station in Kansas City, MO. The station was in need of 'a body' to work a shift. By age 21, his career was developing in the Bay area at KOBY in San Francisco and KEWB in Oakland as Jim Wayne. He landed at KMJ in Fresno after getting fired at KEWB. Ron Jacobs came calling one day and offered him a job at KMAK Fresno where the Jim Price name emerged. He found himself in the middle of the radio wars with KYNO. He arrived at KDEO in January. Jim is credited for breaking the *Mamas and Papas'* "California Dreamin" as a national hit that many stations refused to play.

Upon his arrival, station owner Howard Tullis sat down with Jim and explained that he as a Program Director would cost him money. The sales people make him money. He tasked Jim to get the ratings up so he could sell the station for a good price. He added 'There will be no promotional money and no bonuses if you succeed." Jim then presented the situation to the staff and said "guys, this is the opportunity to see how good we are."

Buzz Baxter (Noon-3 PM)

Robin Scott (3-7 PM)

Larry Boyer (7-11 PM)

Mike Ambrose (11-5:30 AM)

Up to this time, KDEO had been focused on beating KCBQ. But KGB caught on fast as Boss Radio was inaugurated with no commercials during the first weeks of the format and minimal talk thereafter. This created a tremendous setback to KDEO and caused new thinking about the direction the station should take. Jim Price applied the same philosophy he had used at KMAK: When you have low ratings you have fewer ad spots to run. So you just play more music. One strategy put into service was editing records to reduce the run time and increase the number of songs played per hour. With Jim Price contributing DJ talent and ideas, Tom Schaeffer was equally busy as news director and putting Jim's ideas into practice. Tom was even busier as a weekend DJ at KOGO under the name of Tom Sherman. A modest raise brought that to an end.

Line-up:

Fred Kiemel (6-10 AM) arrived after March from KPOI in Honolulu, HI to take this shift when Robin Scott returned to evenings. Fred moved to KMEN San Bernardino in 1966.

Jim Price (10 AM-2 PM) later became a successful station manager at KSDO, KGB, KKBH, and KYXY. Jim retired from radio in 1990, but remains active in Airwatch traffic reporting in San Diego.

KDEO RAYDEO AM 910

Frank Terry (Terrance Crilly) (2-6 PM) arrived to replace Hal Pickens. He was part of the legendary Fresno radio war at KMAK, and was on staff briefly before becoming a Boss Jock at 93 KHJ Los Angeles.

Ray Willes (Ray Willes) (2-6 PM) began his career at KOCO in Salem, OR, but was a seasoned veteran by the time he arrived to KDEO. He was in Omaha in January when he received a job offer to go to WCCO in Minneapolis. He was reluctant to take the job because he was just starting his vacation and he had preferred an area not prone to heavy snow. So he accepted a job from KDEO that emerged at this time. He replaced Frank Terry. While at KDEO, Ray was also on the air at KFMX, a jazz station in La Jolla, under the name of Johnny Desmond.

Robin Scott (6-11 PM) 'Rotten' Robin Scott overcame the disability of polio to develop a successful radio career. He replaced Larry Boyer in this shift before moving to KCBQ in 1966 and later surfaced as Bob Jackson at KBBQ and KLAC in Los Angeles and KSON in San Diego.

Buzz Baxter (11 PM-6AM) replaced Corky Mayberry after his shift was eliminated.

Smiling Lee Shoblom (Weekends) returned from Denver in 1965. After leaving KDEO in 1966, he moved to all-news KBTR Denver, became PD at KRAM Las Vegas, then applied to the FCC for his own station KFWJ in Lake Havasu City, NV in 1970. He added an FM station and TV station that he operated until 1997. He has served as President of the Arizona Broadcasters Association, Director of the National Association of Broadcasters, and was elected to the Broadcasters Hall of Fame in 1999. He now enjoys semi-retirement as a radio

1966
End of Tullis and Hearne Ownership and Top 40 Format

The Boss Radio phenomenon proved to be too much for KDEO. By February, KDEO eliminated its format to do all requests. The station was acquired by Mort Hall, Mort Sidley, and Don Balsamo in late 1965. The ratings had improved to third in the morning drive and second in the afternoons. Despite the improved ratings using this format, they proceeded to experiment with oldies radio, then adult standards. The timing was poor for these formats to succeed, so the station remained adrift under numerous ownership changes over the ensuing years.

As a footnote, seventh grader Tommy Irwin helped out at the station as a gopher for the news department. He would grow up to be Shotgun Tom Kelly.

The final Top 40 lineup was as follows:

Fred Kiemel (5:30-9 AM)

Jim Price (9-Noon) remained on staff until 1968.

Ray Willes (Noon-4 PM) By 1966 the station adopted an all-request format and Ray was moved to overnights. He started 'moonlighting' by doing a morning show at KGIL in LA. When pressed to decide on one job or the other, Ray selected the larger LA market. His career took him to all the major markets in the country. Ray now does voice-over production work for television networks, corporate clients, and motion picture production companies.

Pat Moore (4-8 PM) replaced Robin Scott when he went to KCBQ. Moore had worked at KACY in Oxnard and went to KRLA in summer 1966. His Class 1 license established his expertise as an engineer as well as an air talent. He recently retired from the California Fish and Game Public Information Office.

Buzz Baxter (8-Midnight) Buzz went on to establish Buzzy's Recording Studio in Los Angeles. He has since died.

Lee Shoblom (Midnight to 5:30 AM) was moved to overnights. He recognized that substantial format changes were coming under the new ownership and left again to Denver.

Oldies Radio 1966-1976

The station switched to an all request format in February 1966, followed by oldies later in the year. The 1969 line-up featured the following: **Royce Johnson** (6-10 AM) went to KOGO San Diego in 1970. He died in the mid 1970s. **Ron Reina** (10 AM-2:30 PM); **Gary Seger** (2:30-7 PM); **Mike McGregor** (7-12 PM); and **Stan Schwam** (12-6 AM). **Gary Allyn** arrived in 1973, bringing former KCBQ jocks **Neil Ross** and **K.O. Bayley** to join Mike McGregor, **Rod Page**

KDEO claimed to be the first station to broadcast the *American Top 40* countdown on July 3, 1970.

Album Rock 1976

KDEO adopted an album format in 1976 and became Music 910 Album Rock. The staff included: PD **The Cruiser** (Larry Himmel) (6-10 AM); **Les Tracy** (10 AM-3PM); **Jesse Bullet** (3-7PM) former Boss Jock at KGB; **Ernesto** (7-11PM) had been PD at KPRI San Diego; Classic Album Hour (11PM-12AM); and **Susan Bradley** (Weekends)

KMJC: Adult Contemporary 1977-1979

The station was acquired by Lee Bartell in 1976 and his son Richard was installed as General Manager. Kevin Metheny replaced 'The Cruiser' Larry Himmel as PD. They quickly changed the calls to KMaJiC and adopted the 'Magic 91' slogan. The format was adult contemporary and popular music. Ratings soared high enough to knock KCBQ out of its long-standing market lead, and gave rise to B100 (KFMB-FM) in 1978. Among the DJs were over the years were:

1977: **Steve Goddard** who went to KCBQ in mid 1977, **Kevin O'Brian (Metheny)** who is now Director of Program Operations for Clear Channel, **Tommy Sarmiento** who served as Production Director, Music Director, and had the name Jackson Lee on the air;

1978: **Roger W. Morgan** served as PD and now hosts the syndicated Rock N' Roll Rewind show based in St. Croix in the US Virgin Islands. He also does some acting on motion pictures. His staff included: **Chris Lance, Jackson Lee** and **Chris Cane**, (who would all move to 13K in 1979), **Bobby Malik, Alan Beebe**;

1979: **Jeff Salgo** and **Willy B**.

KMJC: Christian Radio 1980-2002

The calls created an opportunity to redefine the station as KMJC (King and Master Jesus Christ) as religious broadcasting commenced. The station is now owned by Family Radio Network. The calls have since been changed to KECR.

FOOTNOTES

[1] *Borderlands*, An El Paso Community College Local History Project, Rosemary Hoy and Peonie Pompa

Images – Raydeo KDEO AM 910

The 1964 'Riot'

Former KDEO Studio in El Cajon – 2003.

Old 'Mic' Stand on Sign

Album format from 1976.

As KMJC 1977-1979

Tullis & Hearne designs from early and mid 1960s.

Chuck Daugherty broadcasting from a studio remote at the beach.

Design for album cover in 1963.

XEAK Radio AM 690

There are three transmitter tower sets along the highway in Rosarita Beach, Baja California, Mexico. The tower set located most southerly in Rancho Del Mar is XEPRS, (formerly XERB) where Wolfman Jack howled over the airwaves for decades. The others are XEKAM that featured an all-business format where former Boss Jock Johnnie Darin worked. The final one is XETRA that is all-sports. Before it was XETRA, the station was XEAC, renamed XEAK, that offered a Top 40 music format from April 1957 to May 1961. Of these stations, XEAK was the only one with a studio in San Diego.

Under Mexican law, Americans cannot own broadcast stations in Mexico, but they can lease all of the air time. That is the arrangement that Jim and Bob Harmon of Tulsa Oklahoma made with owner George Rivera, who remained a partner, as they took over 5000 watt station XEAC in 1956. Jim became the station General Manager based in San Diego and Bob became Sales Manager based in Hollywood. They boosted the power from 5000 to 50,000 watts in Spring 1958 with an orientation that covered all of Southern California. A new transmitter building was built on the road to Tecate near La Presa, BC and live broadcasts were transmitted daily from 6 AM to 4 PM. A recording studio was located in a two bedroom bungalow behind the station offices at the Mission Valley Inn in San Diego. There, evening DJ's taped their shows for broadcast the following evening. The station calls were changed from XEAC to XEAK to create a new image in 1957. The Harmons came up with the slogan "The Mighty 690". This emphasized the dial location and de-emphasized the Mexico-based station. A third Harmon family member, Helen Harmon Alvarez, was the part owner of Tulsa TV station KOTV that featured programs for children in the afternoons. The sta-

tion offered a basic format by only playing the top 40 songs repeatedly, plus the 'Pick of the Week'.

The 1957 charter air staff consisted principally of DJs that would be prominent in San Diego for decades, but not necessarily under their original names. Operations Manager Bill Brown preferred two or three syllable names for quicker listener identification. So several 'temporary' names came over the airwaves. The line-up was identified as follows:

Jim Lyle (6-9 AM) arrived from WBAC (location unknown) and featured 'Happy Music' in the morning, as described by the station marketing material.

Noel Confer (9-11 AM) was a staff announcer at KOTV in Tulsa, OK and appeared on Jim Ruddle's children's TV show a few times as a space villain. Noel had also done some work at Tulsa's first Top 40 radio. Noel was freelancing in Los Angeles in 1956 when Jim Harmon and other friends visiting from Oklahoma invited him to join  them to watch bullfights in Tijuana. During that outing, Jim invited Noel to join them in their new radio venture. His show was oriented toward housewives. Music focused more on ballads and ads promoted household items.

Bill Ray (11 AM-1 PM) and (7-9 PM) hailed from Canada but arrived at XEAK from KFMB in San Diego. He was asked to join the XEAK venture in "something called Top 40" over 50,000 watts clear channel. He had never heard  of this before and thought it sounded interesting and on the edge, so he came aboard. His day show was aimed to appeal to all types of listeners. His night show, Bill Ray's 690 Serenade, was softer music.

Stan Evans (1-2 PM) discarded a law degree for a radio career.

Artie Lee (2-4:30 PM) served as the afternoon drive DJ and Program Director.

Bud Handy (Ted Porter) (4:30-7 PM) served as the evening drive DJ and as the station's Chief Engineer.

Ted Lake (9-Midnight) had a show featuring dance music aimed at younger listeners.

Noel Confer, Bill Ray, and Ted Lake were the first DJs hired. They were followed shortly by Stan Evans, and then Artie Lee. It was customary then for DJs to announce the news live during each others' shows. The station started producing weekly surveys starting in July 1957.

1958

Bill Brown's experiment using minimal syllable names came to an end when it was learned that personalities broadcasting from a Mexican transmitter into the United States had to use their real name. This produced the names familiar to listeners in the San Diego area for decades to come.

The station produced surveys that originally provided a hit list framed by a blue letterhead at the top and an ad at the bottom. Beginning November 7, the survey sported the Top 40 list framed by the DJ staff in individual music notes. Each DJ offered their personal picks for future hits.

Noel Confer (6-9 AM) replaced Ted Lyle in this shift. Noel used the nickname "Two Gun". He also was the station news editor. His talents have extended to acting on the live stage, in movies, and on television. While doing a play at the Old Globe Theatre he became involved with the assistant director to the dismay of a stage hand. He was doing TV spots at the time as "Two Gun Noel Kelly" and the stage hand asked Noel and his director friend "What kind of nickname is Two Gun?" She immediately replied, "That isn't a nickname. That's a conservative estimate".

Jim Ruddle (9-Noon) While attending the University at Tulsa, Jim Ruddle hosted a childrens' show as a masked pre-sputnik, pre-NASA spaceman who introduced Little  Rascals episodes. Jim left KOTV in 1957 to attempt a transAtlantic sailboat voyage that ended as a result of a mid-passage hurricane. He returned to KOTV for a few months in 1958, then went to XEAK.

He came with the expectation that he would be the full-time newsman. He never expected, intended, nor was overjoyed at being a Top 40 disc jockey. This would be his only radio gig before he became a news anchor at WGN Chicago.

Stan Evans (Noon –2 PM) enjoyed attending bullfights in his past time.

Bill Ray became ***Frank Thompson*** (2 PM - 4:30 PM) and had a trade mark phrase of "there you go.

Ernie Meyers (4:30-7 PM) had a goal in life to become a horse jockey, but he lacked the small stature to be successful and became a disc jockey instead. He got into radio as  an announcer on Armed Forces Radio while serving in the Army during the Korean War. He arrived in San Diego in 1954 and became the morning man at KCBQ. Disliking the Bartell format when they bought the station in 1955, Ernie moved to XEAK in 1958 to assume Bud Handy's shift.

Artie Lee became ***Art Way*** (7PM-9PM) and continued to serve as Program Director. He later went to KOGO, then returned as a DJ.

Ted Lake became ***Bob Donnelly*** (9PM-Midnight) and became Program Director after Art Way went to KOGO.

Glen Boyd (Midnight – 6 AM) was the last to arrive in February 1958 and was the first overnight DJ. He stayed the longest in that shift through April 1959.

Bud Handy left the air but remained on staff as the Chief Engineer.

Confer, Evans, and Ruddle all lived in Baja California and jocked at the transmitter under work permits and visas. Ernie Myers commuted from his home in Imperial Beach near San Diego when he had a morning shift. The old wooden bullring in downtown Tijuana was an immediate draw for many of the Mighty 690 staff who were new to the Mexican culture. The Harmons had used this venue to entertain air personalities they desired to recruit for their station.

Frank Thompson explained that the old wooden bullring in Tijuana was a magnet for many 690 staff. They would be "in the stands on Sundays to glorify the dance of death with loud ole's, along with five thousand others. The sun's shadows crept across the arena and the black bulls died one by one" at the hands of toreros Ramon Tirado and Luis Procuna. Of special interest to the DJ's was Julia, who drove from San Diego each weekend and occupied a poolside room at the La Sierra Motel. She always brought a suitcase full of vodka. She laughed with the matadors' in their lounge, swayed by the pool, and was seated near the flags and the officials of the corrida in the arena. "On Sunday, high combs in her auburn hair held a black mantilla festooned with red rose blossoms. Every torero raised his cap to her following his escabello. She applauded, opened her arms, and threw many kisses as the ole's rained down upon her shining matadors." Her absence was conspicuous as the bullfight season neared its end. Word spread that *"she married someone stateside who had no suit of lights, but did have the power to persuade her to settle down to family life."*

1959

The survey design was in black and white only starting on February 1. Late in the year, the number for the consecutive survey issued was added. DJs often worked in split shifts or short shifts that reduced their time on the air. The line up was:

Ernie Meyers (6-9 AM) took the morning shift when Noel Confer briefly replaced Glen Boyd in the overnight shift. When XEAK dropped their Top 40 format in 1961, Ernie did the morning shift at KOGO radio in San Diego for 19 years that offered a MOR format and non-music programming. He went to KSDO for another 14 years. He went to KPOP in 1997. His career at KPOP and in broadcasting reached a conclusion in November 2000 when he suffered a stroke and was unable to do radio. Throughout this time, he also called races at the Del Mar Racetrack.

Jim Ruddle (9-Noon) left by the end of the year to attend graduate school in Albuquerque, NM and was a University Fellow in the American Studies doctoral program. His career moved toward teaching and doing radio and TV news in Tampa, FL. He became a news anchor at WGN TV Chicago in 1965, then a NBC news anchor in Chicago in 1967 where he remained for 20 years. Upon retirement, he lived aboard a sailboat for four years and now resides in New York.

Stan Evans (Noon –2 PM)

Dick Martin (2-4:30 PM) arrived and left during the year and his shift was absorbed by Stan Evans and Frank Thompson.

Frank Thompson (4:30-8 PM) went to KOGO San Diego working the mid-morning shift. He now lives in his native White Rock, British Columbia, Canada.

XEAK THE MIGHTY 690

Bob Donnelly (8-10 PM) continued as Program Director. He has since died.

Art Way (10-Midnight) By the end of the year, Art was working multiple shifts including the 3-4 PM and 7–9 PM shifts.

Noel Confer (Midnight to 6 AM). He moved to the 9-Noon shift by the end of the year when Jim Ruddle left. Noel retired from radio broadcasting in 1996 and lives in Tulsa, OK.

Len East (10-Midnight) arrived during the year when Art Way took other shifts.

Bob McLaughlin (Midnight – 6AM) arrived during the year when Noel Confer took the 9-Noon shift to replace his friend Jim Ruddle.

1960

The station surveys continued to have DJ photos until January 29. Then surveys stopped production until May 27. When they resumed, they were published on an intermittent basis and there were no references to the DJs, but the number of the survey issue remained. Stan Evans became news director. Listeners recall hearing the jingle 'Wonderful XEAK . . . Swingin' in San Diego . . . Mighty 690 . . . XEAK.'

1961

The station issued their last survey (Issue No. 174) on February 24, at which time XEAK dropped the Top 40 format for oldies Million Dollar Music. On May 6 when Gordon McLendon bought the station, he changed the calls from XEAK to XETRA (as in the late edition of the newspaper) to begin the first 24-hour all news format in radio. Listeners had been asked ahead of time to tune in to the big change on May 6. The day started with a re-enactment of the 1952 World Series followed by the world's first All News Station. Newscasters included Stan Evans, Russ Porterfield, and Bob Bingham. Former KCBQ DJ Earl McRoberts was also on the news staff.

The staff Top 40 remaining at the time of the sale were:

Noel Confer moved to KDEO San Diego radio in 1961 for three years, where he was news director then program director. He later moved on to Country Western stations, including KSON in San Diego, under the name of Two Gun Noel Kelly.

Art Way moved to KOGO, then returned during the year to XEAK. He went back to KOGO after the format change, then moved to KGB in 1963 before it became a Drake Boss Radio station. After that he joined Ernie Meyers at KOGO once again. He was last at KPQP (formerly KGB) during the mid 1980s.

Stan Evans stayed with the station in the all news format. He went to KDAY, Los Angeles in 1966. While there, he did an award-winning documentary on Pearl Harbor. He joined fellow XETRA newsman Russ Porterfield at KFOX from 1969 to 1976. He worked for Albert and James photographers after retiring from radio and died in 1997.

Mike Cline arrived in early 1961 and remained after the conversion to all news, but adopted the more sophisticated Michael Cline name.

My Bucket's Got a Hole In It is a forgotten hit by Rick Nelson that has figured prominently during the final Top 40 days of XEAK. Noel Confer wrote that Chief Engineer Bud Handy was filling in on some taped shows when he announced "Here's Ricky Nelson with *My Hole's Got a Bucket In It.*" Listener Bruce Tognazzini (Ask Tog) contributed his own 1961 listening experience with the following story:

"Mighty 690 ruled the waves with which no US transmitter could hope to compete. The Finest Sound Around had millions of loyal fans that fateful April day, when the song from hell first hit the airwaves. I will never forget turning on the radio that first morning to hear the announcer saying, "and now, Ricky Nelson, with *My Bucket's Got a*

Hole In It." I waited breathlessly for the Irrepressable Ricky, but it was never to be, for out of the radio came a most horrible monotone chant:

'Bompitty bomp bomp de bomp bomp bomp'
Bompitty bomp bomp de bomp bomp bomp
Bompitty bomp bomp de bomp bomp bomp
Bompitty bomp bomp de bomp bomp bomp

When four of these utterances had appeared, I assumed a scratched record. (in the olden days, CD's had actual grooves in them and the laser (actually a needle) could become stuck in the groove, so that . . . never mind).

Suddenly, the lyrics changed dramatically and we heard:

Oh, dee ya dee ya ya de ya ya ya
Ya dee ya ya dee ya ya ya

Followed by an actual English word:

Goin' Bompitty bomp bomp de bomp bomp bomp
Bompitty bomp bomp de bomp bomp bomp

All this time, mind you, the song, with exception of the "Oh" and the "Goin", had consisted of a single A-Flat, repeated over and over.

Just when I thought I must have lost my mind, the verse began once again making all but exclusive use of that wonderful A-flat note:

Well I just got back from outer space
Bompitty bomp bomp de bomp bomp bomp
The chicks out there ain't got no face
Bompitty bomp bomp de bomp bomp bomp
With three pair o' arms and four left feet
Bompitty bomp bomp de bomp bomp bomp
Ooo-ooo
Ooo-ooo
Bompitty bomp bomp de bomp bomp bomp
(Chorus)

To say that I was shaken after the first rendition of this song could be to put it mildly, but this was only the beginning. The announcer pitched a new housing development in Florida, then announced Pat Boone singing *Red Carnation for a Blue Lover*, but instead we heard the graceful strains of:

Bompitty bomp bomp de bomp bomp bomp

Bompitty bomp bomp de bomp bomp bomp

He went on to pitch membership in a tennis club in Connecticut, followed by, "Listen to the Big Bopper with . . . "

Bompitty bomp bomp de bomp bomp bomp
Bompitty bomp bomp de bomp bomp bomp

I couldn't stand to listen to it anymore, but I soon discovered I couldn't stand NOT to listen to it either. I turned back in: "Now Elvis sings his song . . . "

Bompitty bomp bomp de bomp bomp bomp
Bompitty bomp bomp de bomp bomp bomp

I t went on like this for three of the longest days of my and my friends young lives. Never an explanation. Never an illusion by anyone on the air to the demon that had captured our favorite station's very soul.

Soon, everyone in school was listening. Before school, after school, between classes. We were mesmerized, like cobras in a basket. After the first 100 playings or so, we had mastered the complex lyrics and found ourselves, in spite of our hatred for the awful thing, actually chanting along:

Bompitty bomp bomp de bomp bomp bomp
Bompitty bomp bomp de bomp bomp bomp

Exactly 72 hours after the possession began, Mighty 690 suddenly signed off the air forever, taking with it a precious piece of our childhood.

We heard ads for a local clothing company in Cincinnatti, Ohio. We learned that the Fire Department in Schenectady, New York, was starting their annual Toys for Tots drive. But most of all, we learned to sing a perfect A-flat with our eyes wide open and no one home inside."

The song Bruce referred to was titled *Moopity-Mope* by the Boss-tones, an awful song that sounded similar to *Surfin' Bird* by the Trashmen . . . 'A-well-a-everybody's heard about the bird, bird, bird . . . '

In 1968, the station dropped all news on April 2 and changed to Beautiful Music featuring the

XEAK THE MIGHTY 690

sounds of Ray Conniff, Ray Charles Singers, Nat King Cole, and others.

The Top 40 format returned to the Mighty 690 on September 19, 1980. Frank Felix was the mastermind behind the format put into play by Program Director Jeff Hunter and Music Director Jim Richards. The format involved playing a tight list of 10 to 15 songs over and over, grouped among a broader music list. The station took the concept that KCBQ had used in the early 1970s by airing an upbeat and energetic staff that read the same one liners with passion as many as 10 times per hour. The station offered spectacular promotions that gave away cars, houses, and cash awards as high as $1 million. The entire product was aimed to build excitement among listeners. The air staff included:

Jeff Hunter 1980

Jim Richards

Steve Clark 1980

Greg Shannon (Greg Panattoni) worked part-time while jocking full-time at KCBQ as Sonny West. He adopted the name from the Greg Shannon who did the overnight shift at KRLA during the early 1970s, and provided Greg with an early education about radio.

Michael Boss 1981 performed production work as well as do an afternoon shift.

Bob Montague

Sue Delaney

Kris Anderson

Wolfman Jack (Robert Smith) as part of a network syndication.

The 'reborn' Mighty 690 brought about the demise of 13K (KGB). Sonny West, (Greg Shannon) wrote: 'I think XTRA knocked off 13K because it was playing a much tighter playlist. When I was there, the entire library consisted of about 120 records. The power hits rotated every 75 minutes. The entire library would turn over in about 5 1/2 hours time. There was not a lot of fluff with the format. Meanwhile, 13K was doing the "personality" thing with a lot of talk, traffic, news, etc. Mighty 690 just hammered out the hits over and over again.' But listener preferences toward FM stations continued to grow, and Sonny further noted that "I think the powers that be wanted to kick up the demographics a bit so they could (move) away from selling zit creams and teeny bopper products.' The Mighty 690 was the last leading Top 40 AM station in the San Diego market as it faded into another format.

In 1984, XETRA Gold 690 began playing oldies under PD Jim LaMarca. Wolfman Jack was featured on Saturday nights. In 1988, talk radio became the format for about two years.

In 1990 XETRA Sports begins with an all sports format that continues. JACOR purchased the station in 1996 and has acquired other stations since to feed their programming over a broader area. These stations include Santa Barbara 1340, San Bernardino 1350, Lancaster 610, Canyon Country 1220, and other stations. However, beginning in 2003, air staff were based in Los Angeles and use a satellite link to relay to their Rosarita Beach transmitter. With that, the relationship to San Diego was severed.

Images – The Mighty 690 XEAK

XEAK Studio near La Presa BC

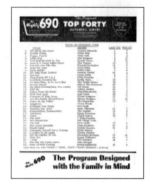

1958 Blue Banner style.

Only two-sided editions November 1958 to February 1959.

February 1959 to January 1960.

May 1960 to February 1961.

Return of Top 40 1982.

XETRA Studio and Transmitter in Rosarita Beach BC, Mexico.

X-TRA All-News format announcement from March 1963.

Coverage Maps

At first glance, coverage maps might appear to be an incomprehensible illustration of technical data, to be valued only by radio frequency engineers. But the maps are a very potent tool to guide decisions about revenue sources, and programming to capture that revenue.

During the earlier days of radio, before the corporate dynasties took over, stations had to compete individually for advertising revenue to fund station operations. As stated earlier, revenue was a direct result of ratings. But the *potential* radio audience was also important to advertisers. Coverage maps graphically represented the potential audience so that advertisers could determine if a station was reaching their desired market area.

The coverage maps further illustrated how antenna orientations and wattage power could dramatically affect the range and number of potential listeners. A comparison of 50,000 watt stations KCBQ and XEAK show how differences in antenna orientations can dramatically alter where the broadcast could be heard. XEAK oriented their antennas toward population centers while KCBQ had to direct their antennas toward the ocean as required by the FCC. Wattage often had to be reduced for evening broadcasts, and the maps illustrate the differences between day and night broadcasts.

Lower powered stations KGB and KDEO oriented their antennas to metropolitan San Diego. This was where audience ratings were collected, so what they lacked in power over a wide area could be compensated by covering a strategic area. This enabled them to compete with the more powerful stations.

Coverage maps provide important information beyond technical data. The maps usually indicate ownerships, network affiliations, and marketing firms that handled advertising accounts for each station. The maps either contained, or were accompanied by, demographic data to aid advertisers in their ad placements.

Coverage maps may also provide artsy depictions using color, geographical features, political boundaries, and station logos. Coverage maps for the four stations featured in this story are provided in the following pages.

COVERAGE MAPS

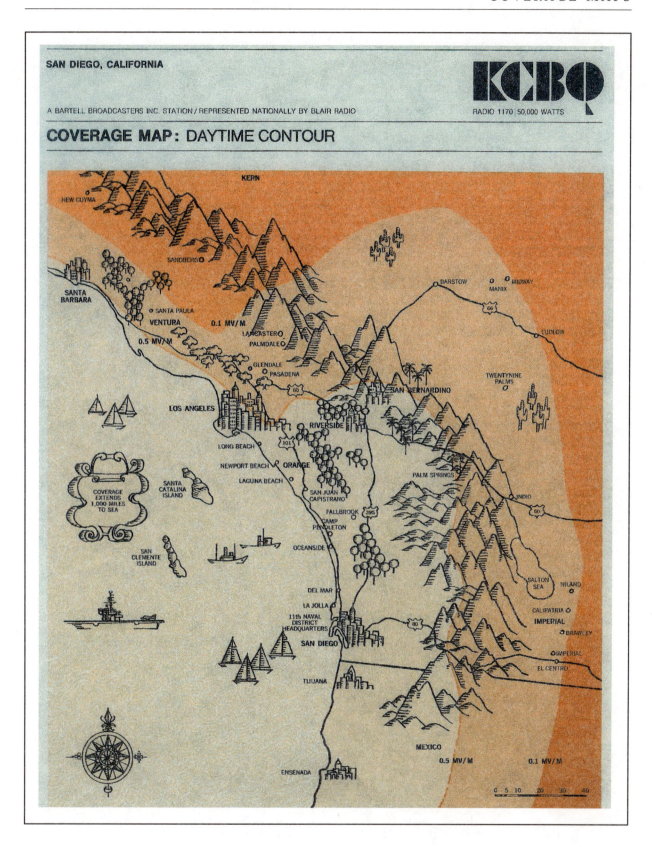

COVERAGE MAPS

COVERAGE MAPS

Sales Office
2566 - 5th Ave.
San Diego, California
Telephone BElmont 4-8144
910 Kilocycles
Center of the Dial

KBAB / KDEO

SAN DIEGO - CALIFORNIA
"Center of the Dial"

Main Studio and Office
200 East Main St.
El Cajon, California
Telephone: Hickory 4-6145
1000 Watts
Unlimited Time

The tremendous growth in population in San Diego since 1950 has raised the city to 20th-ranking among the country's largest cities—with the county ranking 26th in net effective buying income.

The counties within the coverage area of KBAB are among the leading counties in such diversified fields as the manufacture of food products, rubber products, stone, clay and glass products, transportation equipment and printing and publishing.

As a bonus KBAB reaches the 150,000 plus who live over the border—in and around Tia Juana.

NEWS - MUSIC - SPORTS - PLUS FEATURE PROGRAMS BY RCA
MARKET DATA

County	Population	Families	Radio Homes	Grocery Stores	Drug Stores
San Diego	736,196	246,219	239,324	896	184
Orange	222,591	76,492	74,350	332	65
Riverside	63,414	21,209	20,615	121	19
L. A. (Catalina)	1,869	656	638	2	2
Total .5 MV/M Area	1,024,070	344,576	334,927	1,351	270

Population and family figures were derived by applying 1955 Sales Management estimates to the 1950 U.S. Census of Population reports for the minor civil divisions within the indicated coverage area. Radio homes from Advertising Research Foundation's report "National Survey of Radio and Television Sets Associated with U.S. Households May 1954." Grocery and drug store outlets from the 1948 U.S. Census of Business.

Represented Nationally by:
The Bolling Company

Affiliate of:
The Keystone Broadcasting System

COVERAGE MAPS

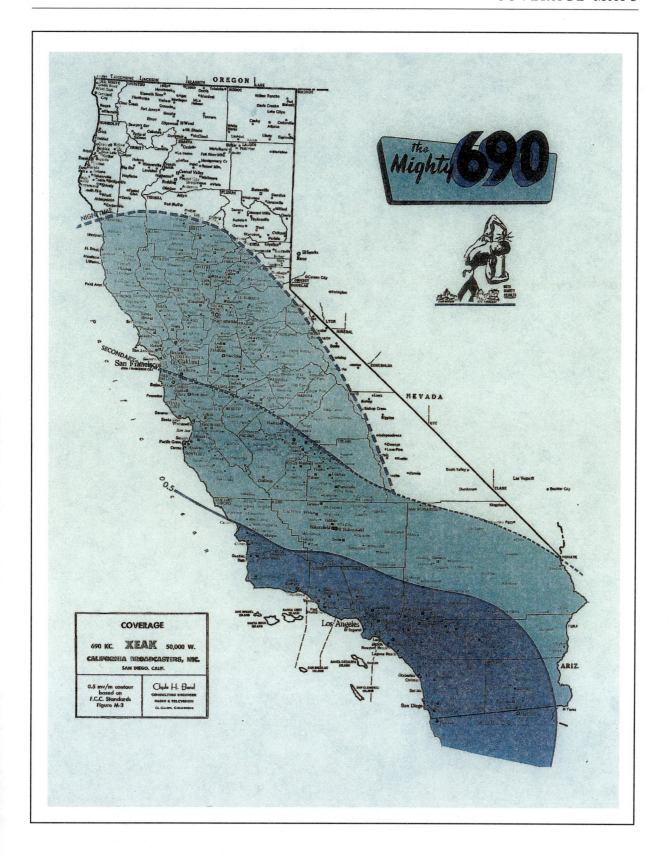

Glossary

Affiliates	Stations that by contract or by ownership align themselves with a network of stations and broadcast the programs produced by that network.
Air Check	Recording of typical broadcast with songs edited to beginning and ending bars only.
AOR	Album Oriented Rock format.
Calls	'Call' Letters identifying station identification. Most broadcast facilities west of the Mississippi River use a 'K' prefix, while most to the east use a 'W' prefix. Stations in Mexico use an 'X' prefix.
Cart	Pre-recorded message on a tape cartridge.
Control Board	Electronic console to modify sound of the broadcast.
Diaries	Print material randomly sent to households to track radio listening habits for the purpose of determining ratings.
DJ	Disc Jockey
FCC	Federal Communication Commission sets rules and standards; and levies penalties involving broadcasting in the United States.
Format	Content of material aired to reach a desired population.
Hourly Clock	Arrangement of material aired over any given hour.
GM	General Manager; oversees all aspects of station operations.
Jingle	Brief music composition and/or narration to identify the station.
Kilocycles	Frequency within the AM band assigned to a station by the FCC.
Market	1. Geographic area a broadcast signal reaches; 2. Population segments a station is programmed to reach.
MOR	Middle of the Road format.
Music Director	The contact person for record company sales representatives to meet and promote records and groups, and who selected songs to appear on the play list. PD often performed this function.
PD	Program Director; manages content of material that is broadcasted
Play list	Songs rotated on a regular basis during a given period of time.
Programming	Pre-planned material for broadcast that is deliberately compiled to reach a particular audience or achieve a particular outcome.
Ratings	Measure of popularity of a station within a given market, and now within a given format.
Rotation	Cycle of songs designed to attract and retain listeners.
Share	Portions of ratings assigned to a station based on audience responses.
Simulcast	Single broadcasts fed over multiple frequencies.
Survey	Station publication, usually published weekly, to list songs on the play list for use by record buyers and to promote a station and its staff.
Top 40	Size and content of a play list that served to define a format, usually aimed at younger audiences.
Transmitter	Broadcast towers with control room to transmit radio signals to receivers.
Tracks	In radio, a song re-recorded on to cartridges as an alternative to a turntable
Watts	Units of broadcast power

Index

Allen Chip 80-83
Allen Dex **27**, 29, 34
Allen Perry 47
Allyn Gary 10, 12, 18, **19**-22, **25**, **29**-34, 42, 87
Ambrose Mike **80**-85
Anderson Kris 96
Anderson Paul Oscar 32, **34**, 35
Azbill WK 5
B Willy 87
Babcock Sam 79-80
Baily Chris 38-41
Balsamo Don 8, 86
Banks Dave 80
Barcroft John 72
Barker Jim 42
Barrett, Don 41
Bartell Family -Radio/Media 1, 3, 11-12, 15, 21
Bartell David 11, 18, 21, 26
Bartell Gerald 11
Bartell Lee 8, 11, 14, 18-22, 26, 44, 87
Bartell Melvin 11
Bartell, Richard 8, 87
Bartell Rosa 11
Baxter Buzz 84, **85**-87
Bayley K.O. (See also Bob Elliott) **34**-35, 58,60, **68**-69, 87
Bayliss John 4
Beebe, Alan 87
Bennett Buzz 10, 32-33, **35**-39, 41, 43, 61, **63**-69
Birrell Harry (See Jerry Walker)
Bishop Bill **17**-19
Bishop Jerry (See Bill Bishop)
Bishop Jerry G. 7, **47**
Boatman Arthur 71
Boles Phil 13, 18, 34 (See also Phil Flowers)
Boss Mike 96
Bowles George 5
Bowman Don 80-82-83
Boyd Barry **23**-25, 27, 29
Boyd Glen **93**
Boyer Larry **85**-86
Bradley Susan 87
Brannigan Alice 5
Brenneman Betty 61
Bright Ben 82-84
Brown Willet 6-7, 29, 64, 69
Brown B Baily **29**
Brown Bill (KGB) **59**-60
Brown, Bill (XEAK) 91-92
Browning Chuck **36**-37, 66, **68**
Bullet Jessie **65**, 68-70, 87
Butts Mile **42**, 44
Cane Christopher 35, **37**-39, 45, **62**-63, 68, **72**-74, 87
Cap'n Billy 70
Carson Jim 27, **59**-63, 65
Carter Gentleman Jim 29-30
Carter Big John **66**-67
Casper Dick 30, 32, 35, 38
Charlie & Harrigan **46**-48
Charter Broadcasting 4, 47
Chenault Gene 6, 9, 55-56, 58-62

Christian Chuck 'Magic' **33**, 35-36, 38-39, 41, 43, 60
Christy Paul (See Johnny Mitchell)
Clark Steve 96
Clay, Tom 83
Clemens Chuck 'Huckleberry' **21**-25, 71
Clifton Jerry 12, 40, 43
Cline Mike 94
Cocker Gary **72**-73
Collins Robert L. **31**-32, 34
Confer Noel 9, **81**-83, **91**-94
Conley Dave **37**, 39-43, 66
Cooper Chuck **57**-58
Cooper Ray 69
Cordell Roy 8, 81
Craig Carol 38
Crane, Jeff 22-25
Crane Marie Breen 5
Crosno, Steve 81
Cruiser (see Larry Himmell)
Dale Don **62**-64, 68
Dandy Coffee Jim **79**-80
Danon Keith 72
Darin Johnnie **60**-61, 91
Daugherty Chuck **83**-85
Davis, Jerry 57-58
Day, Scotty 14, **15**-17, 21-25, 27-28, **31**-**33**, 38
Delaney Sue 96
Dempsey Hoyle 1
Denis Mark **59**-63
Denver Joel 47
Devine Thom **32**, 34
Diego Sam 42
Dixon Mason **45**-46
Donnelly, Bob **92**-94
Dowe Ken **81**-82
Downe Communications 26,41
Downey Morton **84**-85
Drake Bill 6, 9,11, 18, 42, 55-65
Drew Paul 42, 68
Driscoll Jon **73**
Drury Dick **55**
Earl Bill 5, 23
East Len 94
Elliott Bob 58-60, 68
Ernesto 87
Evans Stan 91, **92**-94
Evans Tony 47
Felix Frank 96
Flowers Phil **34**-35,3 7, 48
Foster Bob **64**-65
Fox Don (Sonny) **39**-41
Fox Jimi **43**, 47
Fox Jon 46-**47**
Fox Linda 47
Galore Melody **24**, 27
Gardner Ashley **74**
Gardner Bill 41
Geiger, Chuck **44**-45
Giannoulas, Ted 71
Gilbert Johnny **26**-27
Gillette Rick **72**-74

Gilpin Peri 25
Gladden Ernie 70
Goddard Steve **46**-47, 87
Good David **73**-74
Goss Dean **45**-47, **72**
Green Harold 34
Greenwood Ken 7
Guinn Matt **40**-41
Hall Mel 11, **79**-81
Hall Mort 8, 86
Handy Bud **92**-94
Harmon Bob 8, 91
Harmon Jim 8, 91
Harris, Dick 7
Harris Marion 6
Harrison Mal 31
Hayes Jack **20**-24
Hayes Johnny 57
Harold Drew (See Bobby Noonan)
Heacock Allen 9, 12, 15
Hearne John 8, 79-80, 83-84, 86
Herman Doug 38-41
Hill Jim **30**, 35
Hergonson Bill (See Cap'n Billy)
Herring, Buck 79
Hill Jim **30**, 35
Himmell, Larry 71-71, 87
Holiday Johnny 15, **16**-20, 60, 67
Holliday Sie **79**
Howard Don **13**-17, 33, 74
Huddleston J. Paul 47, 81
Hudson Lord Tim **20**
Hughes Howard 6
Hunter Jeff 96
Hunter, Mike 84-85
Huntley, Chet 3
Irwin Tommy 13-14, 18, 86
   (See also Tom Kelly)
Jack Wolfman 96
Jackson Bob (See Robin Scott)
Jackson Shadoe **16**-17, 19, 33, **79**
Jacobs Ron 6, 23, 67-71, 85
James Ralph **12**-16
Jay Steve **57**-59, 60
Jean-Paul 70
Johnny Bwana **61**, 63
Johnson Batt **37**, 40
Johnson Royce 87
Jones Rex 82-83
Kaiser, Kay 32
Kasem Casey 73
Kaye Barry **65**-66, 69
Kelly Shotgun Tom 35, **37**-38, 41, 43, 45, **66**, 68-**69**, 86
Kelly Gary 42
Kiemel Fred **85**-86
Kirby Jonathon 15
Knight Gary **72**-73
Knight Gene **40**-41, 43-45
LaMarca Jim 96
Lake Ted (See Bob Donnelly)
Lance Chris **72**-74, 87

INDEX

Lander John **71**-73
Lane Lucky 14-15
Langer Stu 25-26
Lee Artie (See Art Way)
Lee Don 5-6
Lee Jackson (see Tammy Sarmiento)
Lee Thomas 6
Lee Tommy **73** (See also Tommy Sarmiento)
Leibert Rich 69-71
Leonard Ron 46
Lewis Fred 56
Light Joe **32**
Limbaugh Rush 24
Linkletter Art 5
London Dave **40**-
Lyle Jim **91**
Mack Jimmy 25-26, **30**-31
MacKinnon Don **80**-81
Maddox Tony 46-47
Malik Bobby 87
Mann Tony **69**
Martin Billy **43**
Martin Dick **93**
Martin Happy Hare 7, **13**-17, 23, **30**, 33, 35, 74
Martinez Danny **39**-40
Marvelle Ted 56
Mayberry Corky **85**-86
Masey Bill **56**
Maule Tom **56**-60
May Peter Huntington 36-38, **65**-66
McCoy Jack **38**-41, 48
McCoy Sherri 38
McDowell Bill 65, 70
McGregor Mike 31, 87
McInnes Jim 71
McKay Bob 73
McKeown Kevin 71
McKinnon Clinton 3
McLaughlin Bob **94**
McLendon Gordon 8, 38, 82, 94
McRoberts Earl **14**, 94
Meeker Jim **58**-59
Mendelson Herb 21, 26
Metheny, Kevin 87
Meyers Ernie **12**, 33, 55, **92**, 94
Michaels Pat 22-24
Millen Ted 79
Minckler Bill 71
Mitchell Gary **58**
Mitchell Gentleman Jim **20**, 23, 56
Mitchell Johnny **67**-68
Mitchell, JB 72
Mitchell Larry **23**-26
Mitchell Lenny **31**, 35, **37**-41
Moffitt Bill 4, **40**-41, 43, 48
Montague Bob 96
Morgan Ray **57**-58
Morgan Roger W 87
Moore Pat **86**
Murphy Tom **19**-22
Murphy Thomas Dr. 63
Murphy Joseph Dr. 64
Nehrbass Mardi 38
Noonan Bobby 40
Ocean Bobby 28, 35, **36**-38, **62**-68

O'Brian Kevin 87
O'Brien Jim **25**
O'Hara Seamus Patrick **17**, 19-20
O'Leary Jim **13**
Pace John 81
Page Rod 87
Paul-Jean (see Johnny Mitchell)
Peterson Gerry 44-45
Pickens Hal **85**-86
Porterfield Russ 94
Prescott Jeff **72**-74
Price Gary 38
Price Sonny Jim 71-71, **85**
Quack Casey B 15, 20
Rabbitt Jimmy 27-**28**
Randall Ted 21
Ray Bill (See Frank Thompson)
Regan Bob 6
Reina Ron 87
Rhine Larry 5
Rich Bobby 37
Richard King 55-**56**
Richards Jim 96
Richards Mark **67**
Richards Stan **19**-21, 25
Richland Tony **64**
Rippey Domino **45**-46
Riddle Sam 79
Rivera George 8, 91
Robbin Rich Brother 12, 35-**37**, 38, **41**, 43, 62, **64**-65, 69
Roberts Brian 45
Roberts Chuck 43-47
Roberts Cliff **74**
Roberts, Phil 17
Rogers Wizard Lew 71
Roman Ruth 8
Ross Neilson **31**, 33, 35, 87
Ruddle Jim 91, **92**-94
Ryan Barry 72-**73**
Sable Leonard 'Len' 26-27, 30
Saint Dick 34, **60**-61
Salek Charles 3
Salgo Jeff 87
Sandin Phyllis 11, 14-15, 18, 21-22, 26
Sarmiento, Tommy (see also Tommy Lee) **44**, 87
Scarborough Harry 34, **37**-38, 48, **65**-66
Schaeffer Tom 85
Schwam Stan 87
Scott Jim (See Stan Walker)
Scott Mike 10, 12, **25**-27, 30, 61
Scott Robin 'Rotten' **22**-23, **84**
Seger Gary 87
Shannon Bob 46
Shannon Greg (See Sonny West)
Sharon Bob 7
Shirley, Ben 15
Shoblom Lee **82**-84, 86-87
Sidley Mort 8,86
Simms Lee 'Babi' **28**, 31- **32**, 34-35
Smith China **34**-35
Smith, Sig 81
Solo Johnny **21**
Spears Michael 42 (See also Mark Richards)

Spencer Raymond E. .80
Stafford Mike 33, 47
Stelljes, Paul (See Johnny Mitchell and Jean Paul)
Stevens Jay (See Steve Jay)
Stone Dave **60**, 62-64
Stone Jay **44**-45
Story Adam 80
Terry Frank **86**
Thomas Ed 82
Thompson Frank 38, **91**-**93**
Thompson Ron 'Ugly' **31**, 34
Tiffany Earl 26
Torgerson Stan 21, 24, 26
Tracy Les 87
Trout Hap 38-40, 42
Tullis Howard 7-8, 79-80 ,82-84, 86
Turpin Les **56**-58-60, 63
Tuna Charlie **39**-40
Van Dyke Charlie **65**-67-68
Vincent Jack **13**-17, 19-27, 34, 38
Wade Bill **55**-61, **82**, 84
Walker Jerry **15**-17, 19-20
Walker Stan 58
Ware Dale 80
Watson Bill 65, 68
Way Art **55**, 91-**92**-94
Wayne Bobby 23-**24**-27
Wayne Ted **74**
Weaver Beau 42
Welch, Digby 70
West Bob **74**
West Gene 36, **61**-63, 65, 67-**68**-69
West Sonny 4, 96
White Brian 43-**44**-45
Willis Ray 86
Williams Dick 80-81
Williams Johnny 19
Williams William F **24**
Wilson George 11, 30, 32, 40, 43
Wisdom Gabriel 80
Wittberger Russ 12,4 1
Wrath John 56
Wright Danny 46
Yale, Arthur 5
Young Dick 41, 43-45

Bold numbers feature photo image